B I R T H Well known to backyard stargazers, the Orion Nebula is a kaleidoscope of gas and dust some 1,500 light-years away. The nebula's intricate, colorful whorls are illuminated by the prodigious light of several hot young stars at its center. In this cauldron of creation, hundreds of stars and planets are now forming.

NATIONAL
GEOGRAPHIC
WASHINGTON, D.C.

EXPLORING
THE
SOLAR
SYSTEM

Other Worlds

J. Kelly Beatty

CONTENTS

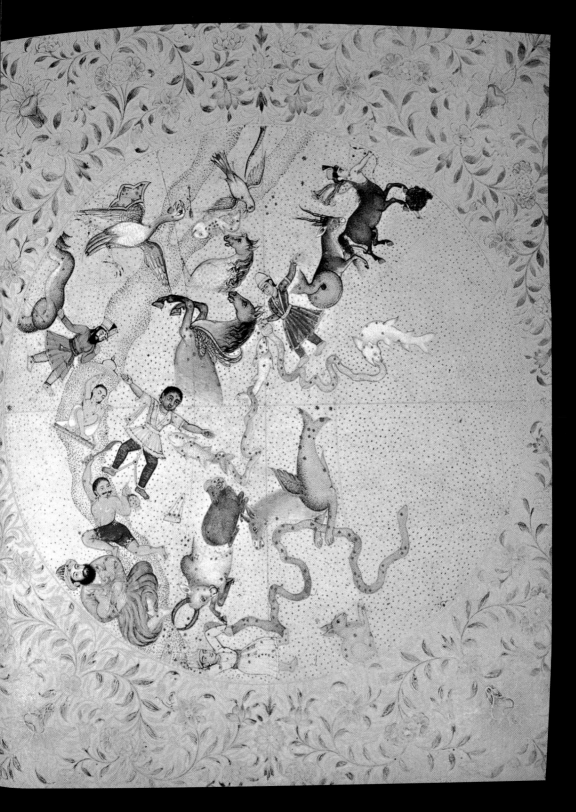

PRECEDING PAGES: Venus and the
Moon grace the dawn sky over
Cuyamaca Lake in California.

ABOVE: This map of the heavens,
circa 1840, was used in creating
a horoscope for an Indian prince.

PRIMEVAL CAULDRON

By the early 20th century astronomers had learned that the universe was expanding rapidly and that it was many billions of years old. A few decades thereafter radio antennas picked up the faint echoes of what came to be called the Big Bang, the spontaneous beginning of everything. These revelations pounded home a sobering, humbling reality: Earth is nothing more than an isolated mote in a vast universe, and we who inhabit it are latecomers to the epic saga of cosmic history. Our home planet may be insignificant in the grand scheme of things, but this corner of the interstellar neighborhood has had an interesting past.

IF WE COULD TIME-TRAVEL BACK TO WHEN THE MILKY WAY GALAXY was about half its present age, it would look much as it does now: an enormous pinwheel, 100,000 light-years across, packed with hundreds of billions of stars. Moving through this "island universe" to a spot about halfway between the center and outer edge, we encounter a vast cloud of gas and dust.

The gas is largely hydrogen, with a little helium and a smattering of other compounds, mimicking the makeup of the universe at large. The dust contains carbon, oxygen, silicon, iron, and a host of other "heavy" elements that did not exist in the universe's first eons. Instead, they were created later by true alchemy —some forged in the internal fires of countless early stars, others debuting as the pressure-cooked cinders of supernovae explosions.

Outwardly the giant cloud seems cold, dark, and completely inert. But deep inside, a subtle stirring motion has begun, as here and there the mixture has clumped into congested knots. Whether spurred by the blast wave from a neighboring supernova or merely from the ebb and flow of migrating matter, these knots continue to coalesce and densify.

A metamorphosis is under way: Material no longer moves randomly but begins to stream toward distinct centers. One of the larger aggregations spins slowly at first, then faster, causing it to flatten into a thin, pancake shape.

The spinning disk continues to scavenge gas and dust from all directions, a furious infall that heats the disk rapidly. Just tens of degrees from absolute zero at the outset, the collapsed cloud has become white-hot in only 100,000 years. Meanwhile, the disk's bulging center compresses itself through self-gravity into a seething protostar that grows denser and hotter by the moment. The core temperature and pressure skyrocket until atoms of hydrogen are forced to fuse together, releasing a flood of pure energy. The Sun is born.

Light and heat stream outward from our neonatal star and into the massive wheel of dust and gas encircling it. Temperatures within this solar nebula vary wildly, from more than 2,000°F at the inner edge to just a few tens of degrees at its outer limit.

Matter within the spinning disk has settled into a distinct midplane, where particles begin to collect into larger solids. Near the young Sun, intense heat permits only a few simple mineral compounds and metals to condense and coagulate. Farther out we encounter a "frost line" near the orbit of soon-to-be Jupiter, beyond which water ice condenses. Somehow the flecks of solid matter manage to stick together after colliding, growing to the size of peas, then baseballs, then boulders.

Jupiter and Saturn begin to take shape, amassing many Earths' worth of rock and ice while sucking in the nebula's dense envelope of hydrogen and helium. They work fast, gobbling up gas with a voracious appetite, and assemble themselves in no more than ten million years (by which time the nebular gas has dissipated).

Perhaps because it is closer in, where the nebula is denser, Jupiter fattens up with runaway growth, sweeping clear a huge swath of the disk. Farther out, the pace of accretion is more sluggish. Uranus and Neptune take longer to fashion their ice-rock cores, a delay that keeps them from accumulating much of the nebular gas before it disappears.

MEANWHILE, THE INNER NEBULA TEEMS WITH COUNTLESS mountain-size chunks of rock, planetesimals that will serve as the building blocks for the rocky, inner planets. Within 20,000 years of the Sun's ignition, these have collected into hundreds of Moon-size bodies.

No one knows why our solar system ended up with the quartet of Mercury, Venus, Earth, and Mars. Most likely, it was by chance—and messy. When today's high-speed computers simulate the countless crashes that these objects endured en route to planethood, invariably some grow larger at the others' expense.

One computer run might yield two or three final masses; another, five or more. But they consistently show that planet-building in the inner solar system is essentially complete within 50 to 100 million years.

Thereafter, leftover planetesimals continue to rain upon them, and sculpting of the inner planets' surfaces finally ends with a frenzy of impacts about 700 million years after the interstellar cloud's collapse.

Its rough construction finished, the Sun's dominion has shaped up nicely, with a suite of rocky worlds close in and an assortment of gas- and ice-dominated giants out beyond the frost line. Yet, despite the common cauldron from which they all emerged, no two planets are closely similar, and each embarks on a unique evolutionary path toward the present day.

FOLLOWING PAGES: The glittering star clouds of Sagittarius point the way toward the heart of our Milky Way galaxy, a stellar city of about 400 billion inhabitants.

OTHER WORLDS

CHANGING PERSPECTIVES

I magine gazing up into a clear and truly dark night sky, your eyes feasting on pinpoints of light from thousands of stars. You make out a few bright beacons that rank among the brightest. You track them from night to night and realize that they drift slowly across the stellar tapestry.

You have discovered the planets.

To the ancient Greeks, planet meant "wanderer," a heavenly interloper that represented a prominent god in mythology. In Ptolemy's view the five planets visible to the unaided eye (Mercury, Venus, Mars, Jupiter, and Saturn) all circled the Earth, as did the Moon and Sun. Beyond lay the unmoving stars. Other cultures devised similar schemes. The universe was orderly, cyclical, divinely perfect in form and function.

Sir William Herschel (1738–1822), an accomplished musician and self-taught astronomer, discovered the planet Uranus from his backyard in 1781. He holds a drawing of the planet and two of its satellites.

Today, thanks to the reach into space afforded by modern telescopes and spacecraft, our perceptions of the cosmos are very different from those of the ancients. We know the planets to be worlds, like the Earth, that formed together with the Sun in a single primeval cauldron more than four and a half billion years ago. The planets may have had a shared origin, but each has proved to be as individual as its mythological namesake.

The renaissance of knowledge about our solar system began with the Sun-centered thinking of Nicolaus Copernicus in the 1510s and Galileo's use of the telescope to vindicate him a century later. Through their efforts, the Sun eventually gained center stage, and the Earth was relegated to a subservient role as one of several attendant worlds. Even after this conceptual revolution, however, the scale of our solar system remained static. As it had for millennia, Saturn still served as the most distant planet, an outpost marking the limit of the Sun's realm. But that would change, as a succession of discoveries swelled the solar system's boundaries to ever greater distances.

A BACKYARD BONANZA. By the mid-18th century, telescopes were no longer curious novelties but tools of serious scientific inquiry that found their way into the hands of many intellectuals. One of them was Friedrich Wilhelm Herschel. Born

OPPOSITE: Astronomers use giant radio telescopes, like this one silhouetted against a star-streaked sky, to probe gas clouds in deep space where stars and planets are forming.

in Hanover, Germany, in 1738 and trained as a musician, Herschel emigrated to England in 1757 and adopted the name William. Herschel methodically observed the heavens with a telescope he had built, using a shiny copper-tin alloy, called speculum, for his instrument's light-gathering mirrors.

On March 13, 1781, Herschel spotted something new and unfamiliar, a star that appeared larger than its neighbors. Returning to that same spot four nights later, Herschel found that it had moved, and he suspected immediately that he had spied a comet. But this interloper had no tail; it later proved to be a planet— the first to be recognized since prehistoric times.

Herschel called the distant world Georgium Sidus ("Georgian Star") in honor of King George III. Later, George awarded Herschel the title of king's astronomer with an annual salary of £200. Eventually, in the tradition of naming planets from Greco-Roman mythology, the planet was called Uranus.

DUELING PREDICTIONS. Seen through a telescope, Uranus is a pale blue dot too small to exhibit any detail. So astronomers of the late 18th and early 19th centuries contented themselves with carefully tracking its orbital motion—tedious work, given that the planet takes 84 years to circle the Sun. Their diligence was rewarded, however, when they realized that Uranus had somehow slipped from its calculated path. These orbital excursions, though small, suggested that some

other massive object, yet farther away, was influencing Uranus's motion across hundreds of millions of miles. Herschel had not been looking for the Sun's seventh planet when he discovered Uranus, but by the 1840s astronomers on both sides of the English Channel had set out to find the eighth.

The hunt for a trans-Uranian body began in earnest in 1843, when a young English theorist named John Couch Adams tackled the problem of Uranus's orbital deviancy and, within two years, deduced a rough position for the unseen planet. Adams tried repeatedly to share his findings with England's astronomical higher-ups, but Sir George Airy, the astronomer royal, was unavailable each of the three times Adams called upon him in 1845.

Meanwhile, the French mathematician Urbain Jean Joseph Le Verrier had also studied Uranus's motion and, not knowing of Adams's work, deduced virtually the same location for the hypothesized planet. Le Verrier likewise had trouble rousing interest in a telescopic search, at least in France. But when he dispatched his prediction to Berlin Observatory in Germany, a search promptly began. On the night of September 23, 1846, astronomers Johann Galle and Heinrich d'Arrest used Le Verrier's predicted position to find Neptune.

Today we know that our solar system contains a dazzling variety of planets and smaller bodies. To ancient skygazers, it consisted of just the Sun, Moon, and five starlike "wanderers."

The eighth planet's discovery caused the extent of the solar system to lurch outward once again. Like Uranus, Neptune is much larger than the Earth, so it proved easy to spot. (In fact, both worlds are within easy reach of even modest backyard telescopes.) Emboldened by the triumph of their theoretical predictions, astronomers wondered aloud what else might be out there.

Planet X and Beyond. Astronomically speaking, the United States was a late bloomer. However, by the turn of the 20th century a surge of observatory construction had catapulted American astronomers to the forefront of discovery. Among them was Percival Lowell. Remembered for his unwavering belief in the existence of canals on Mars, Lowell built an observatory that today bears his name in the town of Flagstaff in northern Arizona. But Mars was not Lowell's only planetary passion. Lowell eagerly joined the search for one or more worlds that some thought still awaited discovery beyond Neptune. He and his observatory staff searched in vain for a "Planet X" until his death in 1916, at age 61.

Astronomers in Flagstaff did not resume a serious search for Lowell's Planet X until 1929, when they installed a new photographic telescope built expressly for the purpose and hired a 22-year-old amateur astronomer, Clyde Tombaugh, to run it. On the afternoon of February 18, 1930, Tombaugh spotted a faint blip that had shifted barely an eighth of an inch on a pair of photographs taken six days apart. This was no asteroid or comet—it had moved much too slowly among the stars. Walking calmly into the director's office, Tombaugh announced, "Dr. Slipher, I have found your Planet X." And, remarkably, he had.

The new world of Pluto proved to be tiny—smaller than the Moon, in fact— yet its reality was undeniable, meaning that the solar system's frontier had been pushed outward yet again. This cold, dark little world weaves in and out along a wide ellipse that crosses well inside the orbit of Neptune during its 248-year trudge around the Sun. But a sizable tilt in Pluto's orbit means that the two planets can never collide. Pluto averages 3.7 billion miles from the Sun, so distant that light from the Sun takes five and a half hours to arrive and, once there, is only 1/1,600 as bright as sunlight here on Earth. Fittingly, the ninth planet was named Pluto, ruler of the underworld in Roman mythology.

Frankly, astronomers can't yet pinpoint where the Sun's dominion ends and interstellar space begins. In 1992 observers David Jewitt and Jane Luu discovered the first of what may be billions of asteroid-size bodies well beyond Pluto, but even these do not mark the outer limit. Beyond them lies a vast cloud of perhaps a trillion ice balls, bound very tenuously by solar gravity. Today's telescopes cannot hope to glimpse these denizens of the distant deep. But they make their presence known whenever one of them "escapes" and dashes headlong toward the Sun, creating a glorious comet. Theorists suspect that, ultimately, the Sun's "extended family" may stretch to an astounding two or three light-years— thousands of times farther out than Pluto, half the distance to the nearest star.

OPPOSITE: The summer Milky Way glows with the light from untold millions of stars.

JUPITER
& COMPANY

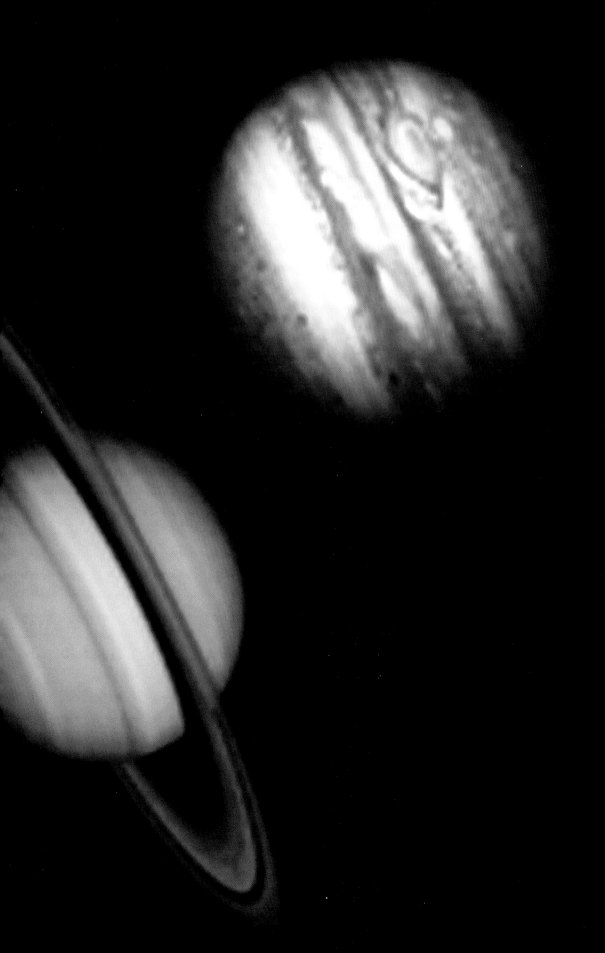

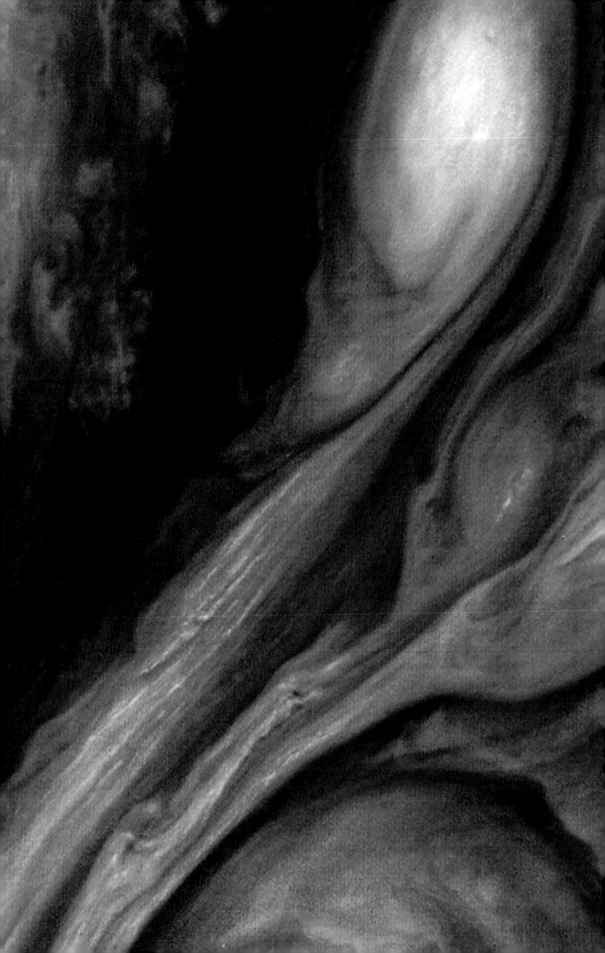

JUPITER & COMPANY

I t is late November, and the still night air has plunged to a bone-chilling 15°F. But the sky above is glorious and crystal clear, with the stars of early winter shining like fiery white diamonds against deep black. High in the east, Jupiter and Saturn gleam side by side, their steady beacons outshining the distant suns around them. Far to the west, too faint to be glimpsed by the naked eye, Uranus and Neptune linger over the horizon. The Sun's outer planets rule the night.

They rule the solar system as well. These four gigantic worlds are sometimes called gas giants because, unlike the rocky inner planets, much of their bulk comes from atoms and molecules of hydrogen, helium, methane (CH_4), and ammonia (NH_3), among others. From the standpoint of composition, the gas giants have more in common with stars than with Earth or Mars. In fact, Jupiter and Saturn mirror the Sun's overall composition rather well: five parts hydrogen, one part helium, and a smattering of everything else.

This makeup tells us that in the outer solar nebula, where temperatures were hundreds of degrees below freezing, planets came together very differently than they did closer in. Most likely, the deep cold allowed rocky and icy matter to solidify and clump into massive cores very quickly; these then attracted great volumes of hydrogen and helium from the nebula. Or perhaps the outer nebula became gravitationally unstable and fragmented into huge rings or blobs that contracted into planets. Either way, our solar system ended up with its four largest members very far from the Sun.

JUPITER, APTLY NAMED FOR THE SUPREME GOD IN ROMAN MYTHOLOGY, is by far the largest planet. With a diameter of nearly 90,000 miles, it is big enough to swallow more than 1,300 Earths. However, Jupiter and its kin do not have solid exteriors; what we see as their "surfaces" are really opaque clouds and hazes suspended within vast, deep atmospheres. If we could plunge into Jupiter's interior, the air around us would soon become compressed into a hot liquid. Farther down, about 6,000 miles below the Jovian cloud tops, atoms and molecules of hydrogen break down into their constituent protons and electrons, creating a molten soup termed metallic hydrogen. The planet's core has a temperature of 30,000°F and a pressure 70 million times greater than the atmospheric pressure at sea level on Earth. While exceedingly high, these are nowhere near the pressure and temperature needed to create the fusion of hydrogen into helium, as occurs in the Sun.

Seen through even a small backyard telescope, Jupiter shows a series of

PRECEDING PAGES: Bluish Neptune and Uranus, ringed Saturn, and huge Jupiter—the gas giants.
OPPOSITE: Tortured clouds race around the rim of Jupiter's Great Red Spot (bottom edge).

light and dark cloud bands running parallel to its equator—our first clue that the Jovian atmosphere is not a static place. In the bright bands (called zones) the air is rising and expanding, causing ammonia and other compounds to condense into bright cloud particles. Jupiter's gravity demands that whatever goes up must come down, which it does in the dark bands (called belts). The sinking air compresses and warms up, which causes the clouds to vaporize. Ferocious winds also blow back and forth across the planet's disk at up to 300 miles per hour.

Jupiter's calling card is its Great Red Spot. Known to telescopic viewers for more than three centuries, this storm is at least twice the size of Earth. It rotates in a counterclockwise direction about every six days, with the wind at its margins racing along at up to 250 miles an hour. We don't yet know what accounts for the spot's color or its longevity— or for that matter what drives the planet's strong winds. One clue is that Jupiter radiates to space about one and a half times the heat it receives from the Sun, so energy must be percolating upward from deep within the planet. Just from watching the clouds' motion, meteorologists can't decide whether sunlight or internal heat is most responsible for energizing the Jovian winds. But the wind pattern has changed little in nearly a hundred years—the weather on Jupiter is easy to predict!

JUPITER

DIAMETER	88,850 MI
MASS	318 X EARTH
ROTATION PERIOD	9.8 HOURS
SURFACE TEMPERATURE	-236°F
REVOLUTION PERIOD	11.9 YEARS

OUR ONE AND ONLY DIRECT EXPLORATION OF JUPITER'S ATMOSPHERE came in December 1995, when an instrumented probe from the Galileo spacecraft rushed headlong into the planet at 29 miles per second. Slowed by atmospheric friction and a parachute, the probe radioed data from seven instruments as it drifted downward. After 58 minutes, the transmitter succumbed to heat and pressure, but the mute probe continued down into the hot depths. Its fate was first to melt, then to evaporate away, yielding its titanium and aluminum to the planet.

Cosmic chemists had predicted that Jupiter would have three distinct cloud layers: crystals of ammonia on top, a middle layer consisting of ammonium hydrosulfide (NH_4SH), and a thick deck of water at the bottom. But Galileo found only wisps of the ammonia and ammonium hydrosulfide, and water was hardly detected at all. As luck would have it, the probe had dropped directly into a partial clearing where the clouds had thinned out. Another surprise was the strength of the Jovian winds, which remained at nearly 400 miles per hour through most of the descent. If the winds are powered by sunlight, they should have died away deep down, so perhaps Jupiter's dynamic atmosphere is energized by heat from within after all.

The planet's interior is pumping out more than heat, because all that slowly churning metallic hydrogen creates a magnetic field powerful enough to extend millions of miles into space. If this magnetic bubble, or magnetosphere, could somehow be made visible in our nighttime sky, it would appear several times larger than the full Moon. Trapped within the field's electromagnetic grip are countless ions and electrons that whip around the planet at tremendous speeds (despite its size, Jupiter rotates in less than 10 hours—and so does its magnetic field). Consequently, the Jovian magnetosphere is a dangerous place, a high-radiation environment that would prove lethal to would-be astronauts.

SATURN, URANUS, AND NEPTUNE. Think of Saturn as a smaller, calmer version of Jupiter. The ringed wonder ranks second to Jupiter in virtually every important planetary statistic: It has a diameter five-sixths as large (74,900 miles), a third of Jupiter's mass, a 45-minute-slower spin rate, and a weaker magnetic field. But at heart Saturn is much like its big brother, with a rock-ice core encased in a hydrogen-helium mix, and it radiates to space 1.8 times the energy it receives from the Sun. As with Jupiter, the energy coming out of Saturn is thought to be a vestige of the planet's hot, furious formation, a lingering trickle of energy released as the interior continues to contract and settle. But that can't quite explain all the heat—something else is going on in there. Perhaps the helium, being denser than hydrogen, is "raining out" toward the core and releasing heat in the process.

S A T U R N

DIAMETER	74,900 MI
MASS	95 X EARTH
ROTATION PERIOD	10.2 HOURS
SURFACE TEMPERATURE	−288°F
REVOLUTION PERIOD	29.4 YEARS

Because Saturn is nearly double the distance of Jupiter from the Sun, it is considerably colder as well. The same triple-layer cloud deck exists in Saturn's upper atmosphere, but the belts and zones are subdued under a cold haze, and there is no Saturnian equivalent to the Great Red Spot. However, this planet does claim one record: the fastest winds. An equator-straddling jet stream is racing eastward at about 1,100 miles per hour—two-thirds of the planet's speed of sound!

Doubling the solar distance once again brings us to the first planet found in modern times. For two centuries after its discovery in 1781, Uranus remained an enigma. Although nearly 32,000 miles across, four times the diameter of Earth, it appears as just a pale-aqua dot. Uranus (and Neptune) did not accumulate vast quantities of hydrogen and helium, as Jupiter and Saturn did. In fact, they are more like "ice giants" than "gas giants," large masses of ice and rock topped with atmospheres a few thousand miles thick. The pale blue hue of Uranus and Neptune comes from abundant methane in their atmospheres, a gas that absorbs much of the red portion of sunlight and scatters away the blue.

One characteristic proved very unusual, however: Uranus does not spin "upright," as the other planets do, more or less. Instead, its polar axis is tilted more than 90° with respect to its orbital plane, giving the impression that the planet is rolling on its side. During its 84-year orbit, Uranus cycles between having one pole pointed toward the Sun, then the other. Astronomers don't know for certain how this situation arose, though presumably the planet suffered a monumental impact when very young that knocked it off kilter.

With so little to go on, planetary scientists had high expectations when the Voyager 2 spacecraft finally reached Uranus in 1986, nine years after its launch. But the planet was a disappointment. At that time only its southern hemisphere was in sunlight and showed virtually no detail, even though the spacecraft came close enough to resolve features only a hundred miles across. It's unclear why Uranus was so lackluster, but astronomers note that virtually no excess heat is escaping from its interior. One surprise from the flyby was that the planet's magnetic axis points nearer to the equator than it does to the poles.

Three years later, Voyager 2 reached Neptune, with very different results. At least twice as much heat is being pumped into the planet's clouds from below as they receive from sunlight, and consequently Neptune is about the same temperature as Uranus despite being much farther from the Sun. Voyager's cameras saw plenty of cloud activity on Neptune, including a large oval—dubbed the Great Dark Spot—partially rimmed by white "cirrus" clouds of methane. Neptune also has a strong jet stream straddling its equator that tops out at roughly 1,300 miles an hour. However, unlike its counterparts on Jupiter and Saturn, this one races westward (opposite the direction of rotation). Another oddity: Neptune's magnetic field is wildly tipped and off center, just like Uranus's.

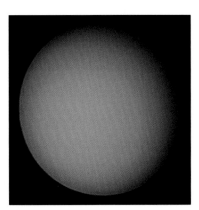

U R A N U S

DIAMETER	31,760 MI
MASS	15 X EARTH
ROTATION PERIOD	18.0 HOURS
SURFACE TEMPERATURE	-353°F
REVOLUTION PERIOD	83.8 YEARS

Fortunately, we are not completely dependent on spacecraft to learn new things about the outer planets. Telescope technology made great strides in the late 20th century, and the astronomer's arsenal was fortified further with the launch of the Hubble Space Telescope. Today we can routinely track the turbulent cloud motions of Jupiter and Saturn, and we have seen giant storms come and go on Neptune. Even Uranus is starting to show signs of activity, as its northern hemisphere begins to bask in sunlight for the first time in decades.

RINGS AND THINGS. When it comes to planets, size must offer certain advantages. In marked contrast to their smaller, rockier siblings, the gas giants are encircled by ring systems, and all four are accompanied by a crowd of satellites. Theorists believe that both attributes are a consequence of how the big planets formed: While growing to final size, they were likely girded by broad, spinning disks—their own, private nebulas—within which numerous smaller bodies materialized. The realization that the outer planets have dozens of moons among them came late in the 20th century. Until the early 1960s, observing the planets was frowned upon by most professional astronomers. Consequently, observing time with the world's best telescopes was rarely squandered on the likes of Mars or Jupiter or Neptune. Gerard P. Kuiper, one of the few professionals willing to buck this trend, discovered a moon circling Uranus (Miranda) in 1948 and another circling Neptune (Nereid) the following year. Four more scattered finds through

1974 brought the inventory of solar-system satellites to 32.

Then the pace of moon-finding really took off, beginning with the discovery of Pluto's Charon in 1978. From the Voyager spacecraft alone, 27 of them turned up in images—3 around Jupiter, 7 around Saturn, 11 around Uranus, and 6 around Neptune. In the 20th century's final four years, astronomers coaxed better performance out of their telescopes and were rewarded with 12 more moonlets. (Curiously, the Hubble Space Telescope has none to its credit.) With few exceptions, these finds are no bigger than 50 miles across.

By mid-2001 the count of planetary satellites stood at 88, with 28 for Jupiter and 30 for Saturn. Most inner moons orbit very near to their host world's equatorial plane and have probably been there from earliest solar-system history. But many of the outlying ones do not. Jupiter, for example, has two quartets of moons clustered in very similar "irregular" (inclined, eccentric) orbits. Each foursome probably resulted from the collisional breakup of a single larger object. Uranus has a clump of like-minded moonlets in irregular orbits as well.

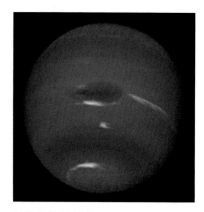

NEPTUNE

DIAMETER	30,780 MI
MASS	17 X EARTH
ROTATION PERIOD	18.8 HOURS
SURFACE TEMPERATURE	-353°F
REVOLUTION PERIOD	163.7 YEARS

We don't usually think of Saturn as a moon magnet. Instead, this planet will always be synonymous with its spectacular ring system. In 1610 Galileo saw odd appendages on either side of Saturn through his crude telescope, concluding that the planet was adorned with a pair of handles that mysteriously disappeared two years later. The great Italian astronomer did not realize that Earth periodically passes through the ring plane as Saturn moves in its orbit. The telescopes got better, and in 1659 Christiaan Huygens deduced the rings' disklike nature.

The next great leap followed nearly two centuries later, in 1857, when the Scottish physicist James Clerk Maxwell demonstrated theoretically that the rings were not solid but instead must consist of countless particles spread in a very thin sheet. "Thin" is an understatement, for the 170,000-mile-wide necklace that is visible telescopically can't be more than about 100 feet thick. By analogy, if Saturn's rings were as thick as this page, the ring system would be a circular sheet of paper a quarter mile across! Although their true composition remains uncertain, the individual particles (which vary in size from dust grains to house-size) must be coated with nearly pure water ice.

A decent backyard telescope will show a dark gap in Saturn's rings called the Cassini Division. But in 1980 Voyager 1's camera revealed a thousand gaps at scales down to ten miles. This is much more structure than current theories can handle. A handful of gaps can be explained by the gravitational influence of nearby satellites, but generally the individual chunks should spread themselves into a smooth, seamless sheet. Perhaps some property prevents the ring particles from acting like a fluid, or large unseen objects within the system may be forcing the ring material into its highly organized structure. The dynamicists who've wrestled to understand theoretically what's going on around Saturn hope the Cassini orbiter, due to arrive there in 2004, will solve this enduring enigma.

The ring system of Uranus is very different from Saturn's: a series of ten narrowly confined ribbons of debris that are very dark and faint. In fact, they were found totally by accident. In March 1977 a team of astronomers had maneuvered a NASA research plane high above the Indian Ocean so that the plane's onboard telescope could monitor a star's occultation behind the tiny disk of Uranus. Unexpectedly, the star blinked out several times both before and after being covered by the planet itself, and the team soon realized that the planet was girded by a nested set of thin rings. Most are only a few miles wide—the fattest one spans 60—with remarkably crisp inner and outer edges. Dynamicists believe such tight confinement can be maintained only if small satellites orbit near the rings' edges, acting like gravitational shepherds to keep the particles from spreading apart. However, careful scrutiny by Voyager 2 turned up just two shepherding moonlets (plus a broad, wispy ring near the planet); presumably there are others, too small or dark to be glimpsed.

Neptune's ring system also came to light through a series of stellar occultations recorded during the 1980s. Astronomers were puzzled that sometimes a star would disappear off to one side of the planet but not on the other side. It seemed that the rings were incomplete or fragmented. Voyager 2's visit in 1989 clarified this confusing situation. Neptune is girded by a series of six faint bands, two of which are quite narrow and the others rather broad. The outer one contains several sections where the ring particles have become concentrated. It's not entirely clear what keeps these partial arcs in place, but that didn't stop astronomers from giving them colorful French names in honor of the bicentennial of the French Revolution: *Fraternité, Egalité, Liberté, and Courage.*

JUPITER MAY BE KING OF THE PLANETS, but its ring system is a poor second to Saturn's. Discovered by Voyager 1 in 1979, the Jovian ring is vanishingly faint and consists almost entirely of particles no larger than those in smoke. Their tiny size makes them very susceptible to a host of forces. Electrostatic charging, for example, creates a "halo" that stays levitated above and below the main ring—and most are driven into the planet within a thousand years or less. Somehow the ring system must be continually replenished with new material, and that source turns out to be a handful of small satellites orbiting near Jupiter. The moonlets are constantly bombarded by microscopic meteoroids, which chip away flecks of matter with every tiny impact.

Rings and moons seem to appear together in the outer solar system, and perhaps with good reason. Both kinds of features may have accompanied each giant planet from the outset. Or rings might represent debris from a satellite's collisional demise. Each system is unique, posing different dilemmas for inquisitive astronomers. The great distances separating us from these graceful, fluid systems make it difficult—and ultimately perhaps impossible—to know exactly what's going on. But that will never prevent us from appreciating their elegant, timeless beauty.

OPPOSITE: Neptune (left) and its moon Triton display slender crescents to a visiting spacecraft.

Galileo, Uranus, and Neptune

History records that on January 7, 1610, Galileo Galilei turned a simple telescope toward Jupiter and discovered a row of three starlike points attending the planet. Within a week he had spotted a fourth. It didn't take long for the brilliant scientist to recognize that these were not stars at all but rather a quartet of moons. However, history also records that not one but two outer planets coincidentally lurked near Jupiter yet remained undiscovered during Galileo's numerous examinations of the Jovian system.

The story behind this astronomical oddity begins with an article in the March 1979 issue of *Sky & Telescope* magazine written by amateur astronomer Steven C. Albers. Based on his computer calculations, Albers determined that Jupiter and Neptune were very close together in the sky around the time of the Italian astronomer's historic observations.

In fact, on January 4, 1613—three years after Galileo first spotted Jupiter's satellites—the disks of the two planets actually overlapped one another as seen from Earth. "It would be extremely interesting," Albers mused, "if Neptune had been mistaken for a moon of Jupiter or an occulted star before it was discovered." Piqued by this very possibility, astronomer Charles T. Kowal and his-

Galileo Galilei (1564–1642)

torian Stillman Drake pored over Galileo's observing notebooks to see if he had noted the interloper. He had! Drawings from the nights of December 28, 1612, and the following January 28th include two *fixa* ("fixed" stars) to the planet's southeast.

Since Galileo rarely included background stars in his drawings of Jupiter, something about these two must have caught his eye.

His notes from January 28th state, "Beyond fixed star [near Jupiter] another followed in the same straight line…which was also observed on the preceding night, but they (then) seemed farther apart from one another." Thus Galileo suspected something but never realized that the second "fixed star" was actually Neptune.

As it turns out, in early 1610 the planet Uranus was *also* in Jupiter's vicinity. Even though Uranus would easily have been bright enough to be seen telescopically, none of Galileo's observing notes indicate that he saw it at all. The two planets were separated by about three degrees, much too far apart to be seen together in the eyepiece. Moreover, says astronomical historian Owen Gingerich, "The magnifying power of Galileo's telescope was too small to see a disk of either Uranus or Neptune, so there was nothing special about these points of light to attract his attention."

Galileo demonstrates his telescope to the councilors of Venice in 1610.

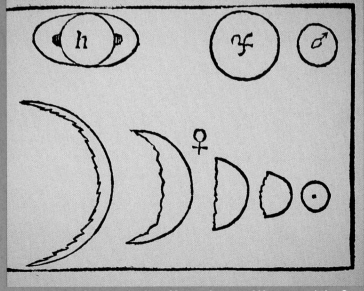
Drawings of Saturn, Jupiter, Mars, phases of Venus, and the Sun.

4 JUPITER

Jupiter (opposite) looms in the camera of the
Cassini spacecraft, which passed nearby in
December 2000. Like the other giant planets,
Jupiter has no solid surface. We see instead
dense cloud layers whose colors arise from
various chemicals. The clouds are stretched into
latitudinal bands by strong east-west winds.

The Great Red Spot (opposite, lower right; and
below, six details) is a hurricane-like storm that
has persisted in the atmosphere of Jupiter for
at least 170 years. Nearly twice the size of Earth,
it rotates in a counterclockwise direction
about every six days with winds up to 250
miles an hour that can gobble up smaller
storms passing nearby.

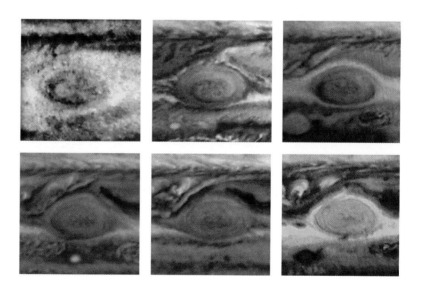

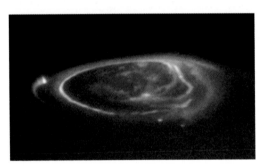

JUPITER When viewed in infrared light (left, upper), Jupiter displays several features not evident in conventional telescopes. The upper atmosphere's warm polar regions glow brightly, as do several cloud bands crossing the globe's midsection. These views also reveal Jupiter's faint ring, whose tiny dust particles have been blasted from the surfaces of small moons circling near the planet. (One of them, Metis, appears as a faint dot tipping the ring just below center.)

An ultraviolet snapshot (left, lower) from the Hubble Space Telescope reveals a glowing lasso of auroral light encircling the polar region of Jupiter. The aurora marks where high-speed electrons from the Jovian magnetosphere are cascading along magnetic field lines into the upper atmosphere. Molecules of hydrogen gas, energized by the electron rain, must give off light in order to cool down. Jupiter's aurora is the most potent of its kind, involving 10 to 100 trillion watts of energy.

Future robotic spacecraft floating in Jupiter's upper atmosphere might see something like this computerized image (opposite). A clearing separates a grayish upper-level haze from the dense, undulating cloud deck below. False colors denote temperature, with purple indicating a hot spot.

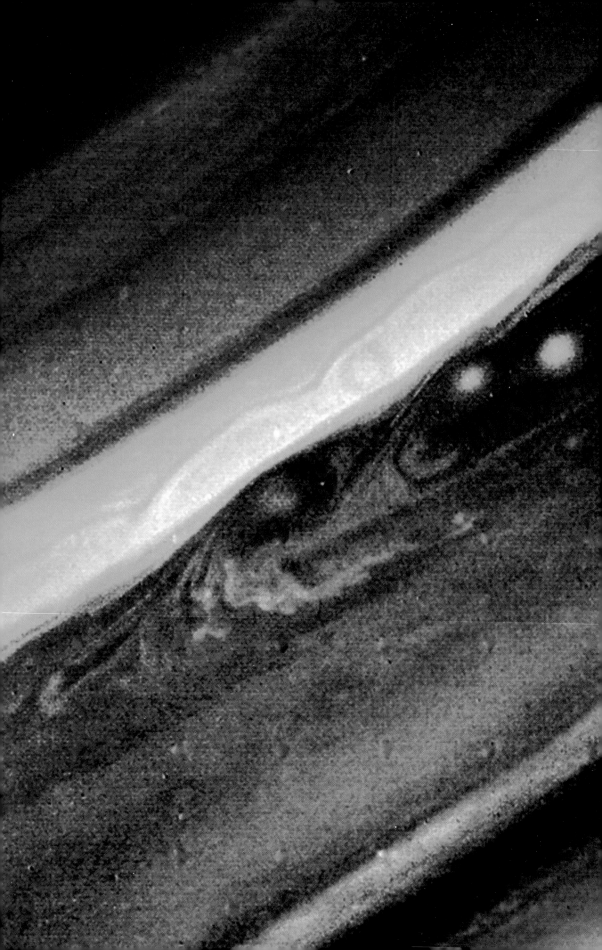

♄ SATURN

Astronomers used the Hubble Space Telescope to record Saturn in August 1995, when the planet's rings were nearly edge-on as seen from Earth. The large moon Titan appears just above the razor-thin rings at left, while its ink-black shadow crosses the planet's disk (shown in false color). Like Jupiter, Saturn has a hydrogen-dominated atmosphere in which rapid jet streams have sheared the cloud layers into east-west bands. However, because Saturn is the colder world, its main cloud decks are lower down and its banding muted by more haze.

When seen by Voyager 2 in 1981 (opposite), the clouds of Saturn exhibited bands and whorls reminiscent of Jupiter's. In natural color, the blue oval in the center would be brown and the two spots below and to its right would be white. The wavy light-blue band is racing past them toward upper right at roughly 300 miles per hour.

FOLLOWING PAGES: Four days after sweeping past Saturn in 1980, Voyager 1 looked back at a dramatically lit crescent encircled by rings.

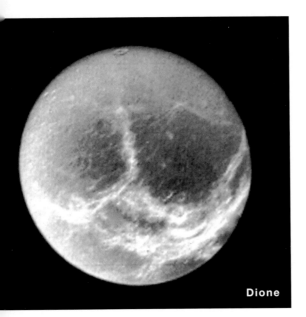

Dione

SATURN   Thanks to visits by the twin Voyager spacecraft, we now see that each of the icy, mid-size moons orbiting Saturn has distinct characteristics. Dione (left, 700 miles across) is crisscrossed by a pattern of bright, wispy markings that may be frost deposits. Planetary scientists suspect that Dione still stirs with geologic activity, because much of its exterior shows a dearth of impact craters—as if it were flooded by slushy ice at some time in the distant past. Perhaps the moon is being heated by tidal energy drawn from its relationship with Enceladus (far right, 310 miles across). Each time Dione goes around Saturn, Enceladus circles twice—a permanent, repetitive resonance that is probably heating the interiors of both worlds through tidal friction. Several types of icy terrain, with widely varying ages, appear on Enceladus. For example, note the smooth-topped "tongue" extending into the heavily cratered region at lower right.

Few worlds in the solar system are as strange as Saturn's two-faced satellite, Iapetus (900 miles across). Despite centuries of study, astronomers still are unsure why one half is pitch black and the other icy white. The most plausible explanations hold that the dark coating came from space rather than erupting from within. Crater-pocked Mimas (250 miles across) bears a huge impact scar named for Herschel, who discovered this moon eight years after spotting Uranus. Tethys (660 miles across) is circum- scribed by a system of long, deep fractures, one of which is 60 miles wide, that probably formed billions of years ago during a titanic impact.

Mimas

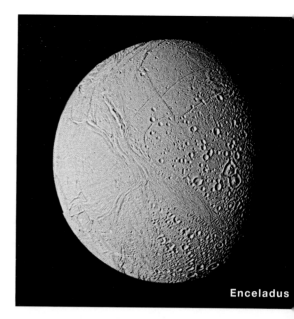

Enceladus

Iapetus

Tethys

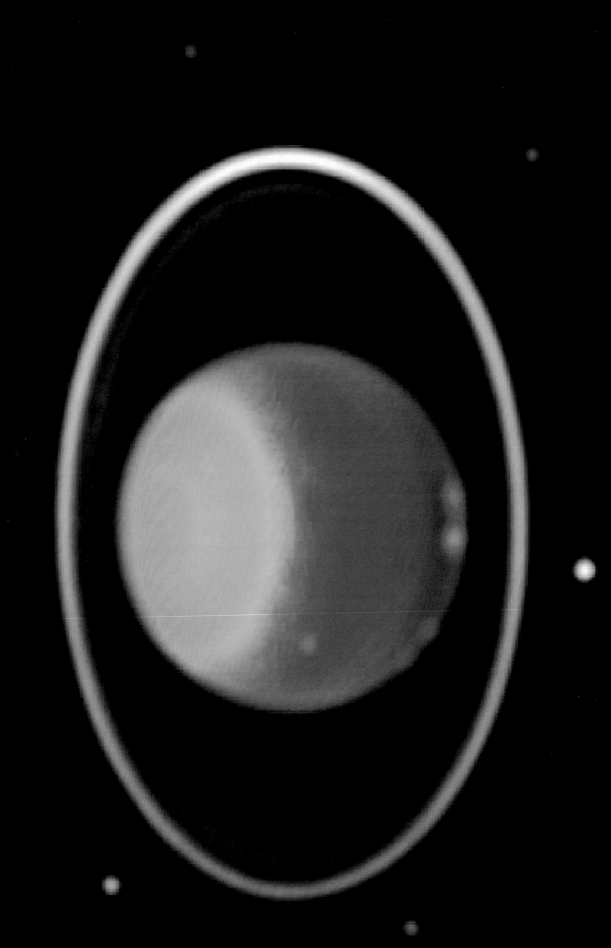

♅ URANUS

From Earth's perspective, Uranus seems to be spinning on its side, rolling its way around the Sun every 84 years. The planet's infrared appearance (opposite) gives clues to the state of its atmosphere, with blue hues revealing the deepest levels and reds denoting high haze. Uranus is girded by a series of narrow rings, as well as a coterie of small satellites named (clockwise from upper right) Belinda, Portia, Bianca (next to ring), Cressida, Juliet, Desdemona, and Rosalind (not shown: Puck).

Images taken by the Voyager 2 spacecraft in 1986 (below) show the bluish cast that Uranus has to the human eye and, to its right, a variation using false color and extreme contrast to bring out subtle details in the polar region. The dark, red-hued polar "cap" is surrounded by a series of progressively lighter concentric bands. Some scientists have suggested that the poles are mantled by brownish haze or smog, which has become arranged into bands by zonal motions of the upper atmosphere. (The bright yellow rim is an artifact of the image enhancement.)

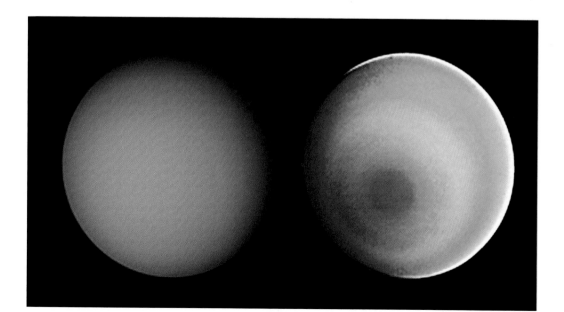

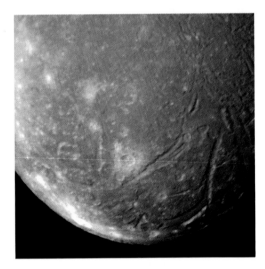

U R A N U S Rivaling Saturn's bands in their complexity, the ring system of Uranus (opposite) was discovered accidentally in 1977 by a team of airborne astronomers. Their data showed the planet to be girded by at least nine rings, the brightest and outermost of which was designated Epsilon. Closer scrutiny by the Voyager 2 spacecraft revealed the system's myriad fainter bands.

One look at the tortured surface of Miranda (left, upper) tells volumes about this 290-mile-wide moon's complex and undeniably violent past. A bewildering variety of fractures, grooves, and variegated regions strongly suggest that Miranda was smashed to pieces as it orbited Uranus— then reassembled itself from the leftover pieces. Although the Uranian satellite Ariel (left, lower) has a diameter of only about 720 miles, it has clearly been geologically active in the past. Most of the moon's southern hemisphere has been intensely cratered and frequently torn by huge fault scarps.

All the satellites of Uranus—and many features on their surfaces—have been named for Shakespearean characters. For example, a map of portions of Oberon (below) highlights a large, dark-floored crater named Hamlet. Unlike some of its satellite siblings, 945-mile-wide Oberon appears to have had a simple geologic history.

Ψ | N E P T U N E

Ever since its discovery in 1846, astronomers have wanted to know more about Neptune. Their wish was granted in 1989, when Voyager 2 culminated its historic outer-planet mission with a flyby of this big, blue world. Since then advances in telescope technology have allowed Earth-based observers to monitor changes in Neptune's atmosphere. In late 1994 the Hubble Space Telescope captured a series of bright clouds (right) in visible and infrared light.

Thanks to the detailed reconnaissance provided by Voyager 2, we now realize that these bright clouds are very high in the planet's atmosphere. The air lower down contains methane gas, which absorbs red light but reflects blue light—a spectral preference that gives Neptune and Uranus their distinctive color. Clouds at very high altitudes reflect sunlight without interference from methane gas and thus look white. Based on the shadows they cast, the delicate, cirrus-like clouds in the weather front seen on the opposite page are some 30 miles higher than the featureless blue cloud layer beneath them.

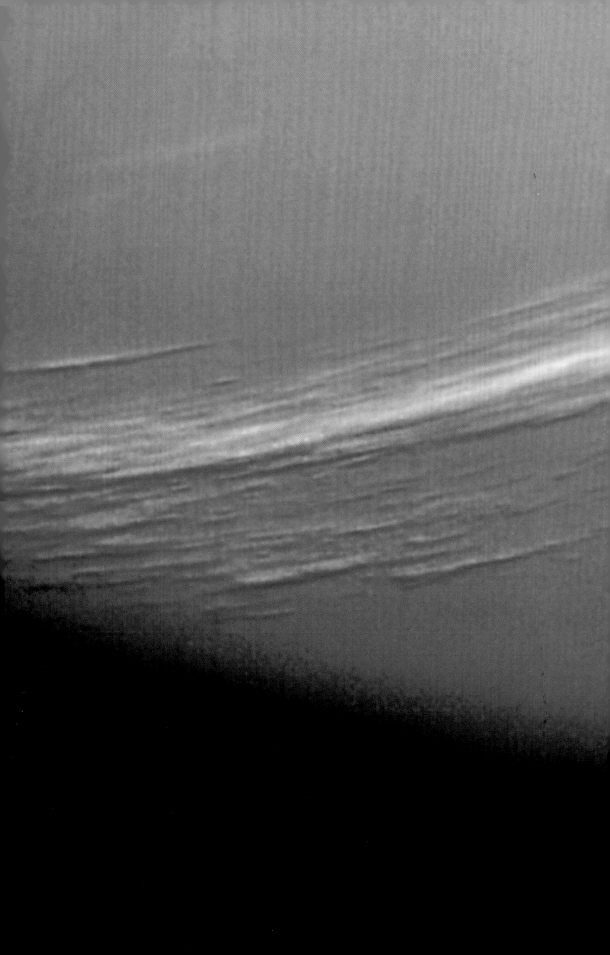

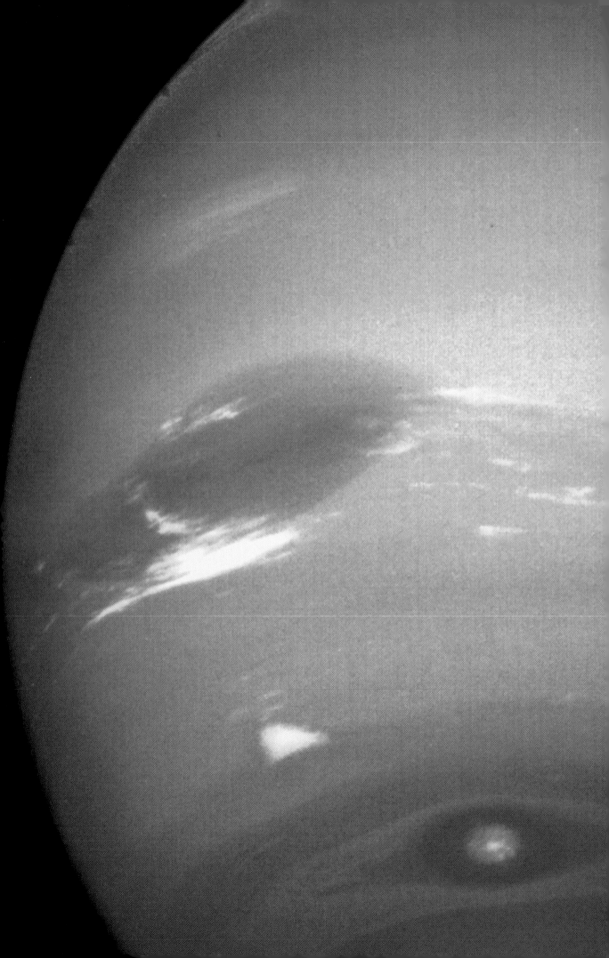

NEPTUNE Planetary scientists were elated by the dynamic atmosphere that greeted Voyager 2 upon its arrival at Neptune in August 1989. The most prominent feature was a huge oval storm, dubbed the Great Dark Spot, that bore many similarities to its counterpart on Jupiter. One striking difference was the bright, white clouds along its rim, whose appearance changed constantly during the flyby.

To the south of the Great Dark Spot was a bright, isolated feature that Voyager scientists nick-named "Scooter." Another cloud system, desig-nated Dark Spot 2, appears near the bottom and has a bright core. Each feature moved eastward at a different velocity, carried along by the planet's strong winds. Curiously, a decade later astronomers could find no trace of any of these features—though numerous others had taken their place.

THE SEVEN
SQUIRES

Pluto and Its Moon

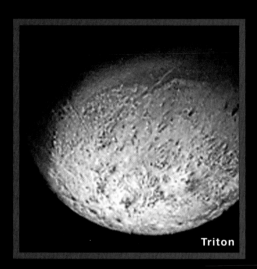

Triton

Europa

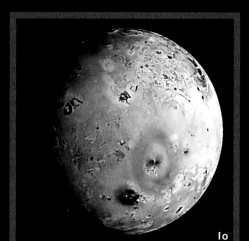

Io

Callisto

Titan

Ganymede

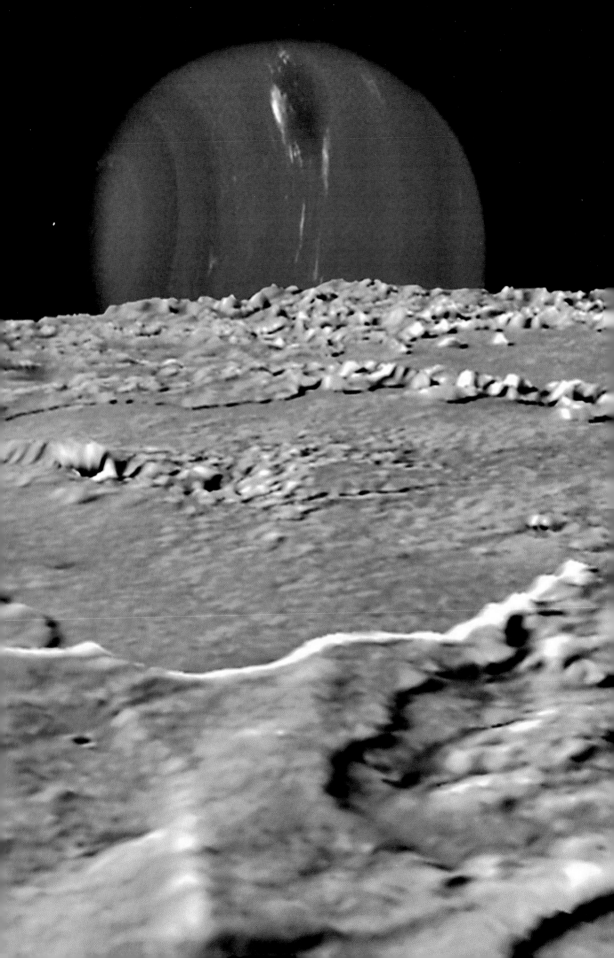

THE SEVEN SQUIRES

Astronomers' quest to find a ninth planet in the early 20th century was driven by the seeming certainty that some large object in the outer solar system, besides Neptune, was affecting the motion of Uranus. Percival Lowell estimated this "Planet X" to have seven times the mass of Earth, whereas William H. Pickering pegged his putative perturber, "Planet O," at twice Earth's mass. And when Pluto's discovery was announced on March 13, 1930—on what would have been Lowell's 75th birthday—newspaper accounts hailed yet another triumph of celestial theory and prediction.

Yet, concerns about the size of Pluto began to surface almost immediately. The new planet was less than half as bright as expected, and it showed no obvious disk even in the best telescopes. Either Pluto had an unusually dark surface, or it was disappointingly small. Then there was its very odd orbit, elliptical enough to overlap Neptune's and canted 17° with respect to the paths of Earth and most other planets. Some astronomers took this to mean that Clyde Tombaugh had actually discovered a giant comet or wayward asteroid. "Naturally, the observatory championed the idea that this was Lowell's fulfillment," Tombaugh later recalled, and despite the initial misgivings it soon became accepted as the ninth planet.

PLUTO'S EXTREME DISTANCE MADE IT DIFFICULT TO STUDY even with Palomar's 200-inch telescope, so for three decades most astronomers continued to assume that the mysterious, far-flung world was roughly comparable to Earth in size and mass. However, dimensional doubts arose again in the 1960s, culminating in 1976 with the discovery that Pluto is coated with frozen methane (CH_4). Since a frost-coated surface should gleam brightly even in weak sunlight, the planet would not need to be large at all to be seen from Earth.

Another demotion followed in 1978, when astronomer James Christy noticed that highly magnified images of Pluto showed a bump, sometimes protruding to one side, sometimes the other, betraying the existence of a sizable moon. Christy named it Charon, partly for the mythical Greek god who ferried dead souls across the River Styx, and partly for his wife, Charlene. The new satellite was a great boon to planetary scientists, who used the size and period of its orbit to calculate that Pluto had very little mass—far less than that of Earth's Moon. So this small outpost could not possibly have enough gravity to attract Uranus in any measurable way. In hindsight, therefore, finding the ninth planet had resulted more from Tombaugh's observing skill than Lowell's predictive prowess.

Pluto continues to reveal its secrets slowly. A fortuitous alignment during

PRECEDING PAGES: Officially a planet, Pluto is sized more like the outer solar system's largest moons. OPPOSITE: Neptune looms over Triton's icy landscape in a computer simulation.

the 1980s caused Charon to pass in front of its parent planet, then behind it, dozens of times. Astronomers monitored these passages closely, eventually deducing that Pluto is 1,430 miles across and Charon 780. The little planet and its big moon, separated by only 12,200 miles, are so close that they twirl together in what is called synchronous rotation: One side of Pluto always faces Charon, and vice versa. Today we know that both bodies remain incredibly cold throughout their 248-year circuit around the Sun, never exceeding -365°F, and they probably consist of rock and ice in roughly equal proportions. Besides methane, frosts of water, nitrogen, and carbon monoxide (CO) have been detected on Pluto's surface, with a wispy nitrogen-rich atmosphere hovering overhead.

From 1979 to 1999, Pluto was nearer to the Sun than Neptune, giving it temporary claim to being the eighth planet. (The tilt of its orbit makes it impossible for the two worlds to collide.) But now, with Pluto edging farther from the Sun each year, astronomers suspect that the deepening cold will soon force the tenuous atmosphere to condense completely as frost. Before that happens, they hope to dispatch a spacecraft to visit the Pluto-Charon system by the year 2020.

Pluto's small size, comet-like composition, and oddball orbit have caused many solar-system specialists to question whether it might more properly be classified as a minor planet, along with the asteroids and other small objects that circle the Sun. Although this identity crisis is hardly new, the debate became particularly rancorous in 1999, two years after Tombaugh's death. The dilemma for the astronomical higher-ups is that no formal criteria exist for what constitutes a planet. Furthermore, Pluto had been called a planet for seven decades, even though it is smaller than several planetary moons. In the end, the International Astronomical Union, the body that sets policy for astronomers worldwide, sided with tradition by decreeing that Pluto's status as the ninth planet should remain unchanged.

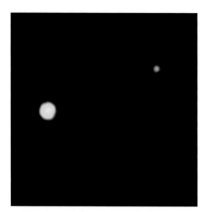

PLUTO

DIAMETER	1,430 MI
MASS	0.2 X MOON
ROTATION PERIOD	6.4 DAYS
SURFACE TEMPERATURE	-388°F
REVOLUTION PERIOD	248.0 YEARS

CHARON

DIAMETER	780 MI
MASS	0.02 X MOON
ROTATION PERIOD	6.4 DAYS
SURFACE TEMPERATURE	-388°F
REVOLUTION PERIOD	6.4 DAYS

TRITON: THE OTHER PLUTO. Part of the confusion over planetary pedigrees comes from Triton, the large moon of Neptune. Discovered by William Lassell in 1846 just 17 days after the planet itself, Triton seemed to be a sizable object from the outset. When Voyager 2 whizzed through the Neptunian system in August 1989, its camera gauged Triton to be 1,685 miles across—smaller than Earth's Moon, but larger than Pluto.

As Voyager 2 drew closer, Triton loomed larger and grew more strange. Vast tracts of the icy surface bore an uncanny resemblance to the dimpled skin of a cantaloupe (today it's still known as "cantaloupe terrain"). Elsewhere were long grooves with raised edges, smooth-topped plains, and patches of fresh-looking frost. Yet something was missing: Triton was virtually crater free. The stunned Voyager scientists had grown used to seeing outer-planet satellites ravaged by eons of bombardment from comets and other large chunks of interplanetary debris. Triton has so few craters, they realized, not because it alone escaped being hit, but because its exterior has somehow been overrun—flooded—within the last 100 million years.

It's hard to imagine how a place so far from the Sun and so unimaginably cold (-390°F, just 70° above absolute zero) could have been so completely awash. At Neptune's distance, the Sun shines with only 1/900 the strength it has here on Earth, far too weak to melt the frozen water, methane, nitrogen, and other ices in Triton's crust. Instead, planetary geologists believe, the overcoating came from within, a deluge of cryogenic "lava" in the form of slushy ices. What triggered so much activity in the geologically recent past remains a mystery, because Triton's rock-and-ice interior should have frozen billions of years ago.

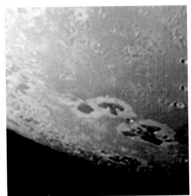

TRITON

DIAMETER	1,680 MI
MASS	0.3 X MOON
ROTATION PERIOD	5.9 DAYS
SURFACE TEMPERATURE	-391°F
REVOLUTION PERIOD	5.9 DAYS

Voyager's images provide some important clues, however. A thin haze seen eight miles up betrays the presence of a wispy atmosphere of nitrogen gas. Clustered around the south pole, and radiating away from it, are more than 100 enigmatic streaks. They look like powder burns, smudges of dark dust that were spurted skyward from the interior, then blown downwind by a gentle polar breeze.

Remarkably, the spacecraft caught a few of these smoking guns shooting their dark plumes miles high into the sky. Yet the gas-powered geysers probably don't arise from stirrings deep within Triton. Instead, weak sunlight may be warming the polar ice below the surface through a kind of greenhouse effect, raising the temperature just enough for bubbles of nitrogen to accumulate under pressure, then explode through a weak spot in the ice cap.

Triton's odd orbit makes all this possible. The moon travels around Neptune backward, or more properly retrograde, counter to the direction of most other satellites. Dynamicists believe that this orbital geometry, in concert with the considerable tilt of Neptune's poles, creates a 688-year-long cycle that produces dramatic swings in climate for Triton. Right now the big moon is experiencing "extreme summer," with the Sun shining on its south pole more directly than at any time since the year 1350.

Conventional astronomical wisdom argues that satellites formed in their planet's equatorial plane, but something clearly went awry with Triton. Over the years theorists have tried to link Triton and Pluto to a common genesis, often resorting to bizarre, physics-defying scenarios. One once popular idea had both worlds encircling Neptune before a chance encounter with some unidentified interloper reversed Triton's motion and sent Pluto careering off into space. But computer simulations prove that Neptune and Pluto have never met (and never will). More likely, Triton started out orbiting the Sun but ventured too near Neptune and collided with one of its satellites. The impact arrested enough of Triton's forward progress to allow its capture by Neptune.

Still, Triton and Pluto may have something of a shared past after all. Their similarity in size and composition is striking—and may not be purely coincidental. Perhaps they once reigned together as kings of the Kuiper Belt, the swarm of icy miniworlds that hover beyond Neptune's orbit. By chance, Triton became enslaved by Neptune, leaving Pluto to serve as the solar system's most distant planetary outpost.

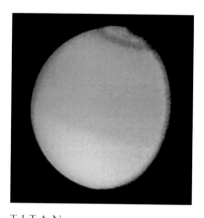

TITAN

DIAMETER	3,200 MI
MASS	1.8 X MOON
ROTATION PERIOD	15.9 DAYS
SURFACE TEMPERATURE	-290°F
REVOLUTION PERIOD	15.9 DAYS

THE TEMPTATIONS OF TITAN. Dutch observer Christiaan Huygens first spied Saturn's largest satellite, Titan, in 1655, and it's been stumping astronomers ever since. Despite being 3,200 miles across—bigger than Mercury—Titan is too far away to show anything more than a tiny dot, using even the best telescopes. One break came in the 1940s when Gerard Kuiper detected methane gas surrounding the big moon, but almost everything else known about it was guesswork.

Ultimately it took close scrutiny by NASA's Voyager spacecraft to learn something of what makes Titan tick. Like the delicate wisps of gas surrounding Pluto and Triton, the big moon's atmosphere is mostly nitrogen—but, unlike those, there's a lot of it. Titan's air mass is six times denser than Earth's at ground level and choked with enough haze to hide the surface from view. The haze is really a photochemical smog akin to the palls that hang over many major cities, a complex mix of hydrocarbon compounds forged through sunlight-driven reactions.

The similarities to Earth don't end there. In fact, Titan probably boasts a robust weather system, complete with wind, clouds, and even rain. Beneath the haze, billowing methane clouds hang in the cold, orange-tinged sky. Cosmic chemists believe that droplets of ethane (C_2H_6) drizzle continuously onto the deep-frozen ground below. They calculate that enough ethane has been wrung from the Titanian atmosphere to form a miles-deep layer over the entire surface—a flammable ocean! (It can't burn because there is no free oxygen.)

But Titan is not completely awash with liquid hydrocarbons. Radar soundings show that something solid must be poking above sea level, and astronomers using infrared-sensitive detectors have glimpsed coarse surface features through the curtain of haze. Titan has at least one large, bright landmass straddling its equator. Perhaps the ethane rain has washed the higher elevations free of dark goo, exposing the underlying crust of white ice. If so, the runoff probably collects into large, widespread seas underlain by a thick layer of organic sediment.

After three and a half centuries of wondering what Titan is really like, real answers are close at hand. Even now one of NASA's most talented robotic emissaries, the Cassini spacecraft, en route to Saturn, has bolted to its side a smooth-sided pod containing Huygens, built separately by the European Space Agency. The combined craft will reach Saturn in July 2004, when Cassini will fire its braking rocket and slip into orbit to begin its close-up scrutiny of the ringed planet.

Huygens's final destination is not Saturn, however, but Titan. Five months after arrival, the pod will surge to electronic life, separate from Cassini, and drop toward the moon. The aerodynamically shaped outer shell will protect Huygens as it slams into Titan's hazy upper atmosphere, then a parachute will slow it further. The spacecraft should descend slowly for two and a half hours, its instruments whirring with pre-programmed activity all the way down. Using Cassini as a radio relay, waiting scientists will be vicariously watching, touching, and tasting the gaseous mix.

Will Huygens land with a thud or a splash? No one knows, though the craft is equipped for either outcome. A miniature laboratory will analyze whatever the craft lands on—or in. Aided by a spotlight, a camera will record the landscape. Weather sensors will report the temperature, pressure, and wind speed. Since Titan's atmosphere lacks oxygen, it may represent a primitive evolutionary state akin to the infant Earth, with organic compounds constantly forming. However, the surface temperature is too cold,-320°F, for the chemicals to react very fast. "Since Titan lacks liquid water, we cannot expect to find examples of the famous 'primordial soup' from which life is thought to have arisen on Earth," explains researcher Tobias Owen. "But Titan can at least supply 'primordial ice cream,' and that may be enough to give us some useful insights."

Space exploration is all about discovery, after all, and Huygens's date with Titan promises to be one of the most anticipated missions in decades.

CALLISTO'S DUSTY MYSTERIES. While Saturn and Neptune each have one large moon that verges on planethood, Jupiter boasts four of them: Io, Europa, Ganymede, and Callisto. Today they are known as the Galilean satellites, though credit for their discovery in 1610 is shared by Galileo Galilei (who recognized them to be Jovian satellites) and Simon Marius (who named them). Like a solar system in miniature, these four worlds display remarkable diversity. Their varied character is a consequence of forming at ever greater distances from the giant planet, which in its infancy radiated huge quantities of heat. Io and Europa, closest in, assembled in a warm environment where rocky minerals could solidify easily but water ice could not. Farther out, however, Ganymede and Callisto amassed metal, rock, and ice with equal ease.

When glimpsed by the twin Voyager spacecraft in 1979, Callisto's darkish

CALLISTO

Cutaway view showing possible internal structure of comparable amounts of ice and rock

DIAMETER	2,990 MI
MASS	1.5 X MOON
ROTATION PERIOD	16.7 DAYS
SURFACE TEMPERATURE	-157°F
REVOLUTION PERIOD	16.7 DAYS

face resolved into a bleak landscape pocked from pole to pole with craters of every size. The surface hovers near -160°F, warmer than Titan but still easily cold enough to make any water ice in the crust behave like solid rock. To have accumulated so many impacts, the icy terrain must be truly ancient. In fact, the lack of geologic variety implies that Callisto is probably frozen throughout and has maintained its frozen sameness for billions of years. It became quietly known as the "boring" Galilean satellite, a world too monotonous to attract much scientific attention.

There are a few curiosities, of course. One hemisphere of Callisto is dominated by Valhalla, an enormous impact feature reminiscent of a giant bull's-eye. Valhalla is not a crater in the usual sense because it lacks a circular rim or central pit. Instead, it consists of a bright, flat center about 375 miles across surrounded by concentric sets of low ridges that extend outward for another thousand miles.

Planetary scientists believe Valhalla resulted from a massive collision early in solar-system history, when Callisto's icy crust had frozen but its interior was still warm and slushy. Whatever smashed into this moon penetrated deeply and sent shock waves reverberating outward through the crust like the ripples created by a stone dropped in a pond. The ghost crater at Valhalla's center is the type of feature termed a palimpsest, named for a piece of parchment that was wiped clean for reuse but still retained vestiges of the original writing.

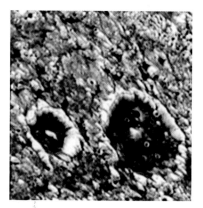

GANYMEDE

DIAMETER	3,270 MI
MASS	2.0 X MOON
ROTATION PERIOD	7.2 DAYS
SURFACE TEMPERATURE	-179°F
REVOLUTION PERIOD	7.2 DAYS

When the spacecraft Galileo arrived on the Jovian scene in 1995, geologists expected that its keener camera would reveal myriad small craters on Callisto's battered surface. Instead, the images show that thick dust coats much of the moon's surface, muting the features lying beneath. In some places tongues of dark debris have streamed downslope and piled up at the bases of ridges and crater rims. Perhaps eons of bombardment have driven off the topmost layer of ice, leaving behind a dusty slag layer. Or perhaps the soil once contained frozen CO_2 (dry ice) that vaporized long ago. Judging from its dark color and its characteristics in infrared light, the mysterious mantling probably consists of clays, minerals, and organic debris. It may be no more than a few feet deep, but that's enough to give Callisto a dark countenance that belies its icy character.

Despite every indication that Callisto languishes in cryogenic quiescence, Galileo found that the moon is not magnetically inert. Something about 60 miles below the surface is interacting with Jupiter's magnetic field, and that "something" is most likely a subterranean ocean at least six miles deep, laced with salts (which make it electrically conductive). Geophysicists are still pondering how such a global ocean could exist without manifesting itself at the surface.

GANYMEDE: A WORLD-CLASS MOON. Of the nearly 80 known satellites in our solar system, Ganymede reigns supreme in size. It has a diameter of 3,275 miles, only a little larger than sibling Callisto, but that extra 10 percent seems to make all the difference. Compared to Callisto's cratered monotony, Ganymede fairly bristles with geologic complexity. Seen from afar, the big moon's face is a study in black and white, carved into large tracts of bright and dark terrain that look superficially like Earth's Moon. However, the similarity ends there: Given the ubiquity of frozen water in this part of the solar system, there's little question that Ganymede's surface is mostly ice. So in the darker regions the ice probably has more contaminants mixed in—whatever those might be—than it does in the bright zones.

In several respects the dark terrain seems reminiscent of the dust-draped icescapes preserved on Callisto. For example, it is almost completely saturated with impact craters, a near-certain indication of geologic seniority. Here and there are palimpsests, the ghostly pale circles that mark ground zero for powerful, primeval collisions. Small, bright knobs and crater rims protrude upward from the vast plains, while dirty-looking piles collect in the valleys between them. "Ganymede's dark material contains abundant clays," explains satellite specialist Robert Pappalardo, "and might crunch under some future astronaut's boot like frozen mud." There's a contribution from organic compounds as well, imported over the eons by countless collisions with comets and asteroids.

EUROPA

DIAMETER	1,940 MI
MASS	0.7 X MOON
ROTATION PERIOD	3.6 DAYS
SURFACE TEMPERATURE	-207°F
REVOLUTION PERIOD	3.6 DAYS

Most of the geologic action on Ganymede takes place in bright regions, where wide, striking swaths of grooved terrain, some hundreds of miles long, crisscross the landscape. These corrugations look as if some enormous rake has been pulled across the surface, in some places more than once and in different directions. The long, parallel sets of grooves and ridges are fault systems, and they tell geologists that parts of Ganymede's crust have been pulled apart by internal forces. As it bulges and stretches, the brittle crust fractures again and again, like a stack of books that shift and slump when their bookends are removed. Close-up images from the Galileo spacecraft reveal this ridge-groove pattern down to scales of just a few hundred yards. But fracturing alone does not explain why the wide swaths are often much brighter than their surroundings. It seems as if some relatively clean fluid—a watery brine or slush, perhaps—has percolated up through the cracks to create the bright coating. No one knows when this resurfacing occurred, but age guesstimates range from four billion years to as recently as a few hundred million years.

Something is definitely stirring beneath Ganymede's icy exterior. In 1996, when Galileo swept past this moon at close range, NASA scientists discovered that Ganymede creates its own magnetic field—a feat unmatched by any other satellite in the solar system (except perhaps Io). The field is stronger at ground level than that of Mercury, Venus, or Mars, and it encapsulates Ganymede in a magnetic bubble more than twice this moon's size. The source may be a molten, churning iron-rich core, and electric currents may also be coursing through a salty, subterranean ocean.

How *did* Ganymede and Callisto become so very different, despite having a shared origin and being near equals in size, mass, and overall composition? One important clue comes from the way mass is distributed within each body, which geophysicists deduced by noting the subtle ways that gravity affected the Galileo spacecraft during its close flybys. If we could slice these moons open, Callisto would show rock and ice mixed throughout its interior, with the rock concentrated toward its center. But Ganymede would reveal three distinct layers: an iron (or iron-sulfur) inner core, a rocky outer core, and a water-ice mantle and crust. This onionlike character means Ganymede was once warm enough for rock and metal to sink toward the moon's center.

Galileo's revelations have left theorists wrestling with a problem of basic physics: To generate a magnetic field, some electrically conductive fluid—be it a molten core or a hidden ocean—must be in turbulent motion. Yet today Ganymede should be frozen through and through. Did a recent pulse of heat jump-start the interior? If so, how? For now, at least, these questions have no clear answers.

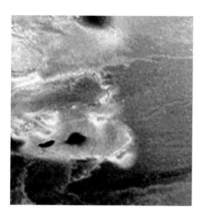

I O

DIAMETER	2,260 MI
MASS	1.2 X MOON
ROTATION PERIOD	1.8 DAYS
SURFACE TEMPERATURE	-216°F
REVOLUTION PERIOD	1.8 DAYS

THE ENIGMA OF EUROPA. When Voyager 1 threaded its way through the Jovian system in March 1979, Europa was the odd moon out. The craft came no closer than 456,000 miles, yet its crude, distant glimpses of Europa still tantalized mission geologists with dark wisps that crisscrossed the frost-covered surface. Voyager 2 ventured much nearer when it arrived three months later, resolving Europa's face into a tangled web of dark streaks, and a few bright ones, that could be traced for up to a couple of thousand miles. Interspersed among them were darkish splotches and mottling that stained the ice like a bad case of mildew.

Amid all this, the science team realized there was something very disquieting about Europa: it was flat, very flat, verging on billiard-ball smooth. Scrutinized down to the images' resolving limit—just a few miles—the surface showed no mountains, no canyons, no great continents or basins, and just a handful of impact craters. It meant that the brittle, icy exterior was only skin deep. Deeper down, instead of having a foundation of rock or rock-hard ice to shoulder the surface relief, Europa had to be soft. It was either a thick layer of warm, pliable ice or, conceivably, a vast subsurface ocean of liquid water.

NASA took 16 years to send its next spacecraft to Jupiter, but Galileo made up for lost time by buzzing Europa at close range 11 times between 1996 and 1999. Europa was still every bit the enigma it had been in Voyager's day. The leading hemisphere, which always faces the direction of orbital motion, is bright and icy. But the trailing site looks darker and mottled, as if the clean ice were mixed with dirt.

And what of the putative ocean? Circumstantial evidence for it abounds. In some places the crust has been cleaved into city-size "rafts" that jostled and tilted one another as they drifted apart, with fresh-looking ice filling the voids left between them. Elsewhere, material has bubbled up from below to form smooth-surfaced ponds. Many of the long dark cracks mimic the fracture pattern that would form if the moon's crust were sliding back and forth in response to Jupiter's tidal pull.

But the strongest argument for a subsurface ocean comes from the deep interior, beyond the reach of Galileo's camera. As with Ganymede, Europa has a gravity field that can be explained by an outer shell of water (frozen or liquid) up to 100 miles thick surrounding a core of rock and metal. When the spacecraft skirted by just 230 miles away in January 2000, it recorded magnetic perturbations emanating from the moon. However, the field is not coming from Europa's core. Instead, Jupiter's powerful magnetic field is inducing a weaker counterpart in an electrically conducting layer not far below the surface—and a salt-seasoned ocean is the best candidate.

Curiously, even as scientists have amassed an impressive circumstantial case for a Europan ocean, they have no direct observations that prove its existence. Nor can they be certain how far down the hypothesized ocean lies: It could be anywhere from a few miles to 20 or more. To know for sure, NASA must follow up with more advanced spacecraft—perhaps an orbiter equipped to measure the amount of the surface's tidal rise and fall, which is proportional to the ice layer's thickness.

Io's Volcanic Wonderland. If Callisto qualifies as the most boring of the four Galilean satellites, Io is without question the most dynamic and colorful. It is a landscape of constant change, a world ruled by dozens of active volcanoes spewing molten rock, sulfur, and superheated gas from its interior. Since the discovery of Io's volcanoes in 1979, some of them have sputtered out, while new hot spots have surged to life. Hundreds more lie quenched or dormant around the globe, their locations betrayed by dark-centered fire pits called calderas.

How can a moon so close to ours in size, 2,260 miles across, wreak so much havoc on itself? The answer is rooted in Io's orbit around Jupiter, and in the special relationship it shares with the orbits of Europa and Ganymede. Jupiter is so massive that all four of its satellites have long since yielded their rotational independence to its powerful gravity. Each is forced to circle the planet and spin in synchrony, such that one hemisphere faces inward toward Jupiter at all times.

In the late 1700s, the French mathematician Pierre-Simon de Laplace determined that the orbital periods of Io, Europa, and Ganymede form a nearly perfect 1:2:4 ratio. This resonant relationship causes Io's orbit to deviate very slightly, just 0.4 percent, from the perfect circle that Jupiter's gravity would otherwise demand. From Io's perspective, Jupiter looms huge in the sky and appears to nod back and forth throughout the 42.5-hour "day." This cyclic rocking creates a repetitive tidal force—manifested as friction and heat—within the moon's crust. The Jovian tide generates so much heat within Io, nearly 100 trillion watts, that widespread internal melting must occur. Volcanoes are the inevitable outcome on Io, and theorists now realize that tidal heating must have a hand in the internal churnings of Europa and Ganymede as well.

A future Io lander, surveying the landscape from ground level, would quickly realize that this is not a hellish place. In fact, well away from the eruptions the ground is brutally cold, about -250°F. The soft, powdery "soil" can take on any number of hues. Most of it is some combination of sulfur, which can appear pale yellow to red to greenish, and sulfur dioxide (SO_2), which looks white when pure but is easily darkened by contaminants. Looking up, the lander would record a delicate mist of spewed particles drifting to the ground, accumulating slowly, steadily, at the rate of about a twelfth of an inch per year. "What this means," explains Bradford A. Smith, who headed the Voyager imaging team, "is that if you were standing around a volcano on Io and wrote your name on a rock one day, it would be covered up the next."

The most impressive of Io's eruptions is Pele, named for the Hawaiian volcano goddess. Visible even from the Hubble Space Telescope, it creates a towering, umbrella-shaped plume some 250 miles high and 800 miles across. No mere lava fountain could attain such incredible heights. Instead, geophysicists believe that a pool of molten rock exceeding 2,500°F has invaded a thick layer of sulfur-laden material, causing a sustained "explosion" of vapor to form the gigantic plume. Condensing in midair as S_2 and SO_2, the fallout surrounds Pele in a wide ring. There the sulfur molecules combine to yield S_3 and S_4, which give the ring its bright red color. Over time these compounds morph once more into S_8, the common yellow form of sulfur.

Despite the ubiquity of sulfur on Io's surface, most of the eruptions are oozing molten rock. Dark rivers of quenched lava are common around many calderas, and occasionally they appear to have melted themselves a channel to flow in as they snake across the landscape. In some spots the crust has ripped open, disgorging fire fountains that gush miles high into the dark sky. At night, these fissures reveal themselves as incandescent ribbons along the caldera rims.

Not all of the volcanic spewings remain on the surface, however. Io orbits closer to Jupiter than the Moon does to Earth, so its surface and wispy SO_2 atmosphere must endure a ceaseless bombardment of charged particles from the Jovian magnetosphere. The onslaught is so intense that roughly one ton of sulfur and oxygen atoms are blasted into space each second, where they soon become ionized (stripped of an electron). Before they disperse throughout the magnetosphere, the lost ions and electrons linger for a time along Io's orbit, forming a doughnut-shaped ring that can be detected from Earth.

Io cannot escape from its orbital resonance with Europa and Ganymede, so Jupiter will continue to inflict its tidal torture indefinitely. But the tons of matter lost to space are a mere dribble compared to the immense reservoir of volcanic "fuel" in Io's interior. So Io will continue to spout off for the foreseeable future, providing this and future generations with a geologic circus that surely ranks as one of the greatest spectacles in the solar system.

OPPOSITE: According to this fanciful portrayal from the back cover of a 1941 magazine, Callisto was a much-sought vacation destination—quite unlike its real-life appeal.

SERENIS, WATER CITY OF CALLISTO

The people of Callisto love beauty, and their city is built on a lake. It is the Venice of the outer worlds. See page 145.

TRITON

William Lassell spotted Triton circling Neptune just a few weeks after the discovery of the planet itself. Although the nature of this large satellite remained a mystery for nearly 150 years, we now recognize that Triton is a unique and fascinating body. Bright patches of frost dot the moon's south polar region, while the dark smudges are probably dingy exhalations from numerous gas-powered geysers that roared to life when exposed to sunlight. Farther north, lingering in the deep shadows and crisscrossed by fractures, is a knobby-textured swath known as "cantaloupe terrain."

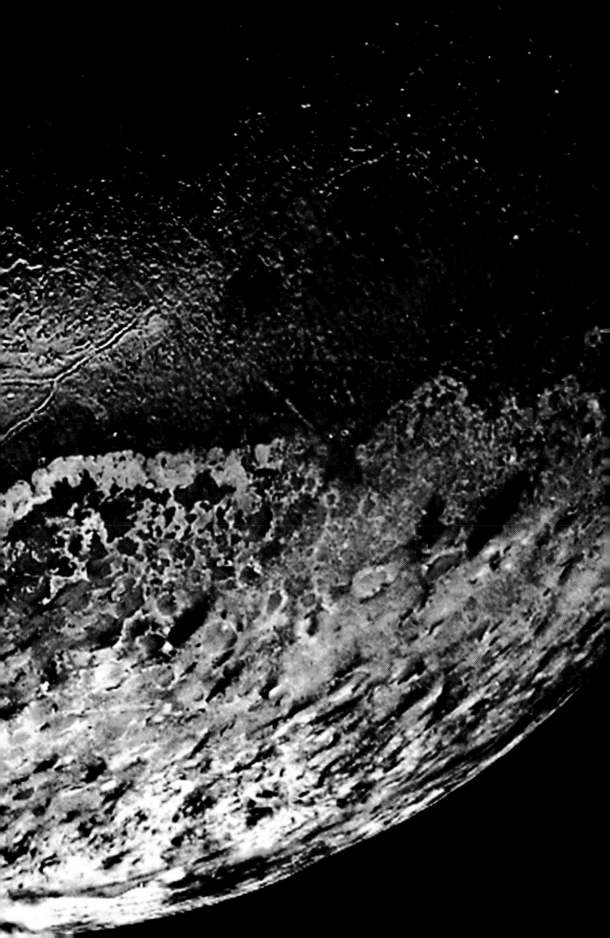

TITAN

The atmosphere of Titan is remarkable as much for its composition (mostly nitrogen, like Earth's) as for its density (a surface pressure one and a half times that at sea level here). Titanian air also contains a little methane, some of which has reacted with sunlight to create an opaque, orange-hued hydrocarbon haze. A false-color closeup of the edge of Titan's disk (left) reveals several layers, tinted blue, floating up to 300 miles above the pumpkin-colored blanket of smog.

If our eyes were sensitive to infrared light, the haze enveloping Titan would seem much more transparent. Infrared snapshots from the Hubble Space Telescope (below) show that the moon's surface is a patchwork of light and dark features. One prominent bright area is 2,500 miles across, about the size of the continent of Australia.

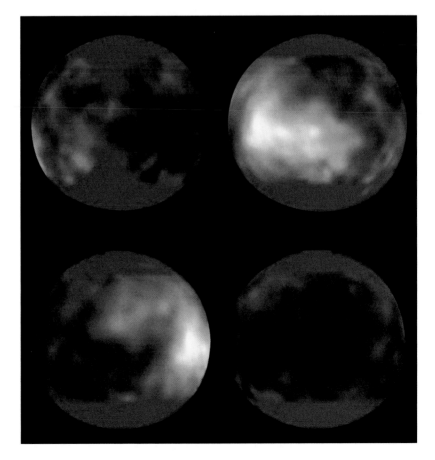

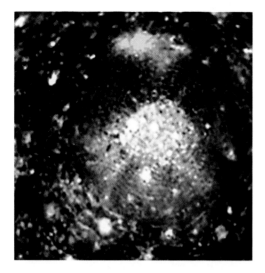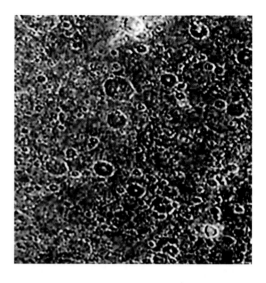

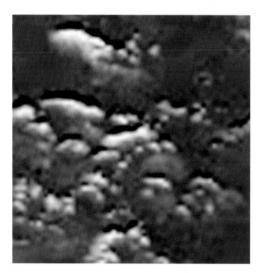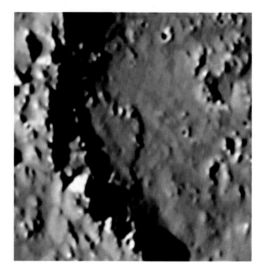

CALLISTO

Ranking third in size among all satellites, Callisto was once considered the least interesting of Jupiter's Galilean satellites. From afar its darkish surface is a largely unbroken sea of impact craters punctuated by a few extra-large scars like Valhalla (upper left), whose 375-mile-wide center is surrounded for another thousand miles outward by multiple rings that look like expanding ripples frozen in place. But Callisto piqued plenty of scientific interest once the Galileo spacecraft arrived during the 1990s. Seen with ten times more detail (upper right), vague bright spots resolved into craters. Another tenfold improvement (lower right) showed that the craters' bright rims were probably crested with relatively clean ice. A final jump to Galileo's most detailed views (lower left) revealed that the surface is remarkably smooth—almost dusty in appearance. Infrared scans (opposite) hint at subtle color differences that are probably related to composition. For example, the giant Valhalla impact apparently punched through the outer crust and allowed cleaner ice (color-coded green) to well up from the moon's interior.

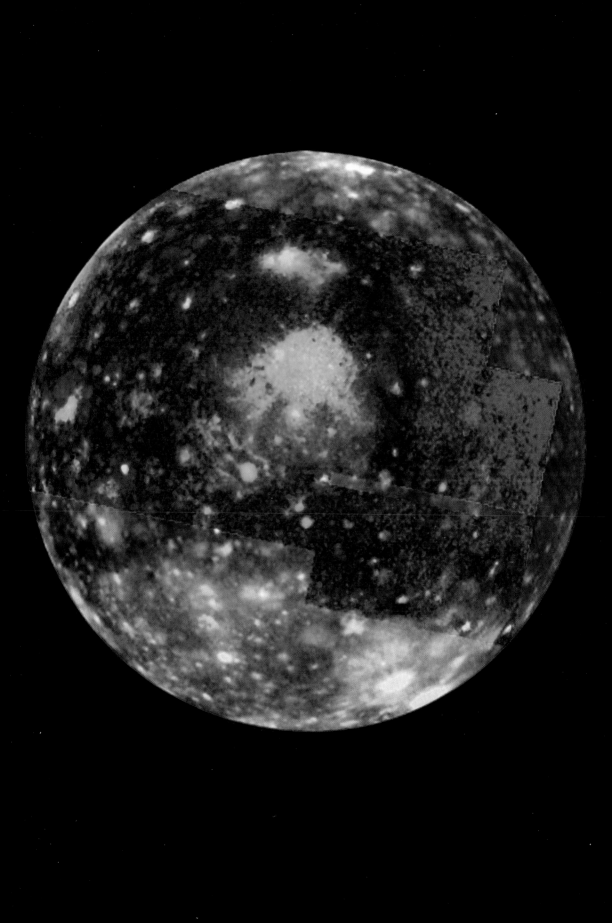

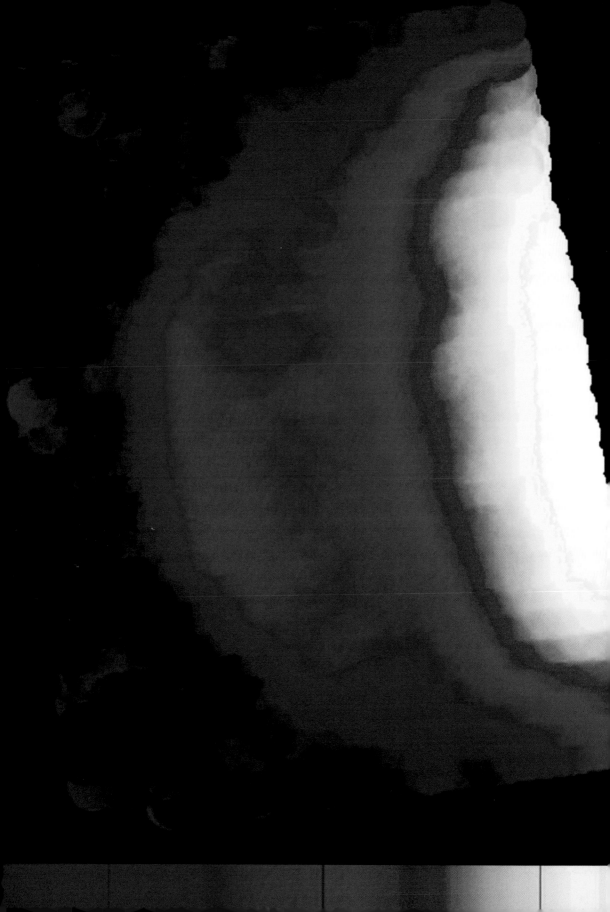

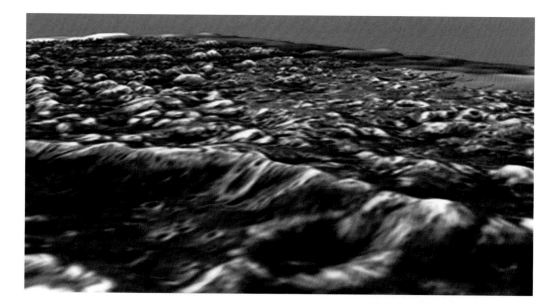

GANYMEDE

When spacecraft images are used to create a three-dimensional portrayal (above), the broad plain on Ganymede known as Galileo Regio becomes a snowboarder's dream, with moguls and furrows stretching to the artificially colored horizon. Based on how it reflects infrared light, the dark material must contain clays and perhaps even some hydrocarbon matter. Among the thousands of craters that pepper this moon is an unusual chain (right) created by a comet that split into 13 pieces before striking.

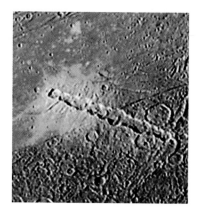

Dawn on Ganymede (left) spreads the Sun's rays across the landscape. But this infrared map shows that even when bathed in morning sunlight (orange and yellow), the moon's surface warms only to about -225°F. Meanwhile, those areas still plunged in darkness (violet, blue) are days away from dawn's light and are 60 to 80 degrees colder still.

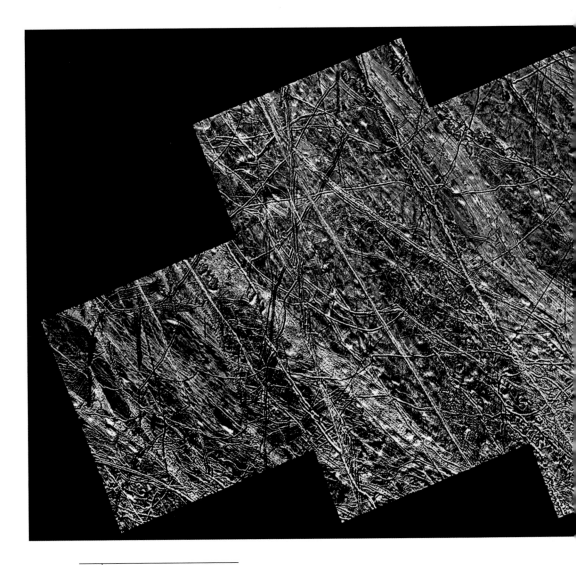

EUROPA

A tract of Europa's northern hemisphere (above), measuring 500 by 220 miles, uses exaggerated color to bring out a web of fractures. Some of the finer breaks are probably due to stress within the crust, caused by the powerful and repetitive tidal pull of Jupiter, just 400,000 miles away. But the larger cracks could mark where water or an ice-water slush erupted on the surface, freezing almost instantly in the extreme cold. Smooth-looking plains, bluish in tone, consist of nearly pure ice and underlie the ridge system. But no one knows what other material contaminates the brownish spots and ridges; they may be a mixture of ice, mineral salts, and organic matter. Very few impact craters pepper this scene because the surface of Europa is very young on geologic timescales.

Most astronomers suspect—but lack ironclad proof—that a global ocean lies not far below Europa's icy crust. Such suspicions were raised when the Voyager spacecraft showed Europa to be almost perfectly smooth and cracked like the shell of a hard-boiled egg (right). These intricate crisscrossing fractures are found nowhere else but on Europa.

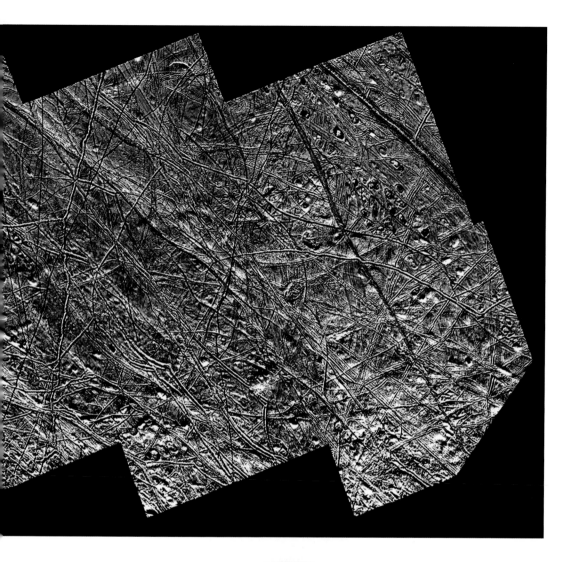

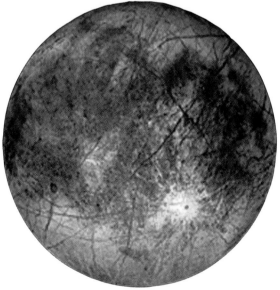

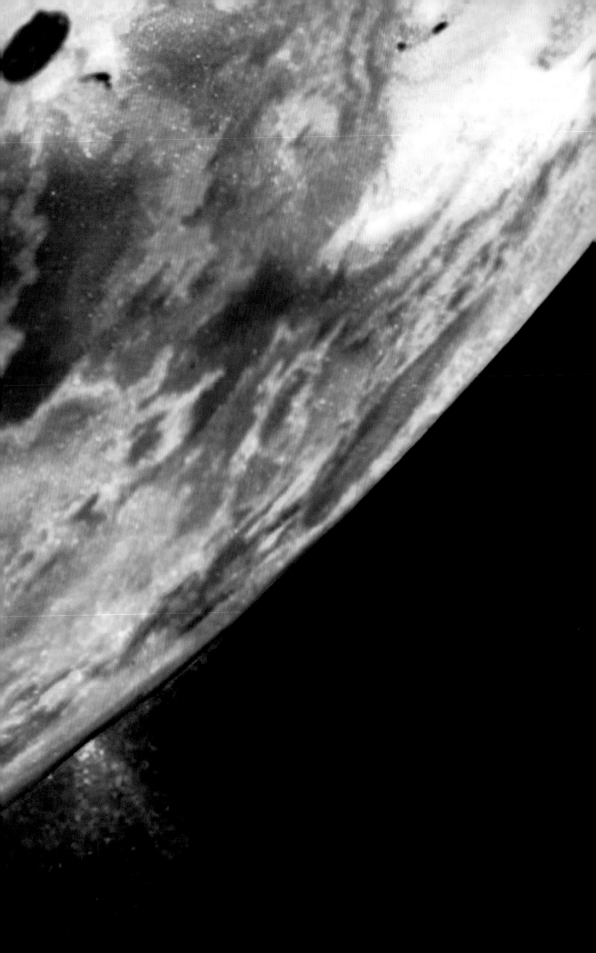

Seeing Is Believing

Sometimes, in science as in life, you have to act on your instincts. When Voyager 1 swept past Jupiter and its suite of moons in March 1979, hundreds of journalists from around the world descended on the Jet Propulsion Laboratory in Pasadena, California, to witness the historic event firsthand. Within days, however, as the hoopla died down, the space center quickly emptied out—even the project's scientists packed up for the weekend.

There was still work to be done, of course. In the mission's navigation section, a young engineer named Linda Morabito (now Linda Kelly) was mopping up the processing of some satellite images taken as the spacecraft headed out of the system. These "optical navigation" frames involved very long camera exposure times, to bring out the background of faint stars that would be used to refine knowledge of the moons' orbital positions. But the images tended to overexpose details on the satellites' surfaces, making them scientifically uninteresting.

Morabito called up a view of Io taken the day before. As she processed it to increase the brightness of dim signals, something caught her eye. Just off the edge of Io was a bright arc looping high over the surface. Someone else might have dismissed it as a furtive artifact, the kind of camera flare that had popped up in many other images. But to Morabito, who had been schooled as an astronomer, this was different. "Right away," she recalls, "I had a feeling that I'd seen something no one ever had before," and she resolved to find out what her "anomaly" really was.

One by one, she ruled out all the possibilities, meticulously recording all the details in a minute-by-minute log. It wasn't a camera flaw after all, nor had another satellite inadvertently slipped into view behind Io. The hours wore on, and the lack of an explanation grew more exasperating.

Finally, it all became amazingly, stunningly clear: Morabito had spotted the towering, 200-mile-high plume from a powerful volcano on Io's surface. A trio of theorists had suggested just weeks before that this moon might show volcanic activity, but something so huge and obvious had caught everyone off guard. "The discoveries were supposed to come from the floor above me," she notes. Within days, the science team found seven more eruptions, all gushing at the same time.

Io turned out to be a volcanic wonderland, and mission planners had enough advance notice to reprogram the camera on Voyager 2, which flew past Jupiter four months later. At least six of the eight volcanoes were still going strong—and most would continue erupting for years to come.

Linda (Morabito) Kelly and her discovery image. OPPOSITE: **A volcanic plume spews from Io.**

IO

When Io slips into Jupiter's shadow (opposite), its many volcanoes put on a dramatic nighttime display. Reds and yellows mark the hottest regions, with several eruptions concentrated on the hemisphere that constantly faces Jupiter, right half. A cloud of hot gas (green) envelops this cluster, as well as the plume over the long-lived volcano Prometheus, left edge. Although this eruption appears only 50 miles high in daylight, the hot gas seen here extends upward some 500 miles.

Caught in the act! Red-hot lava spews along a 40-mile segment of Tvashtar Catena, a nest of interconnected volcanic craters on Io. Two small bright spots indicate where molten rock is exposed at the toes of lava flows, while a dark "L" to the left of the center marks an eruption seen by the Galileo spacecraft several months earlier. Colors have been enhanced to bring out subtle variations.

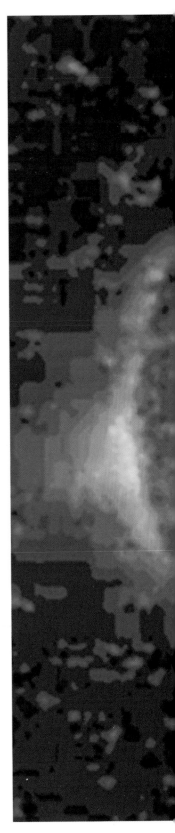

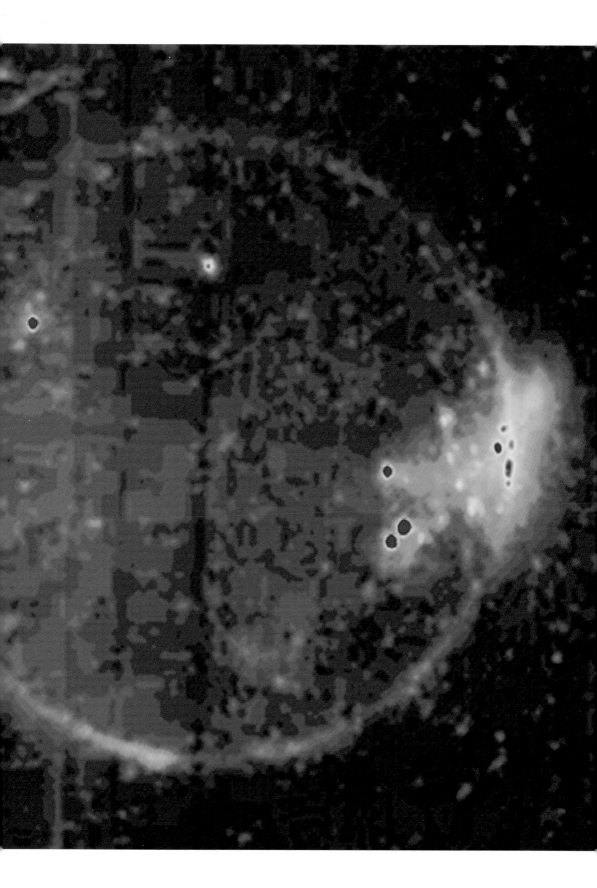

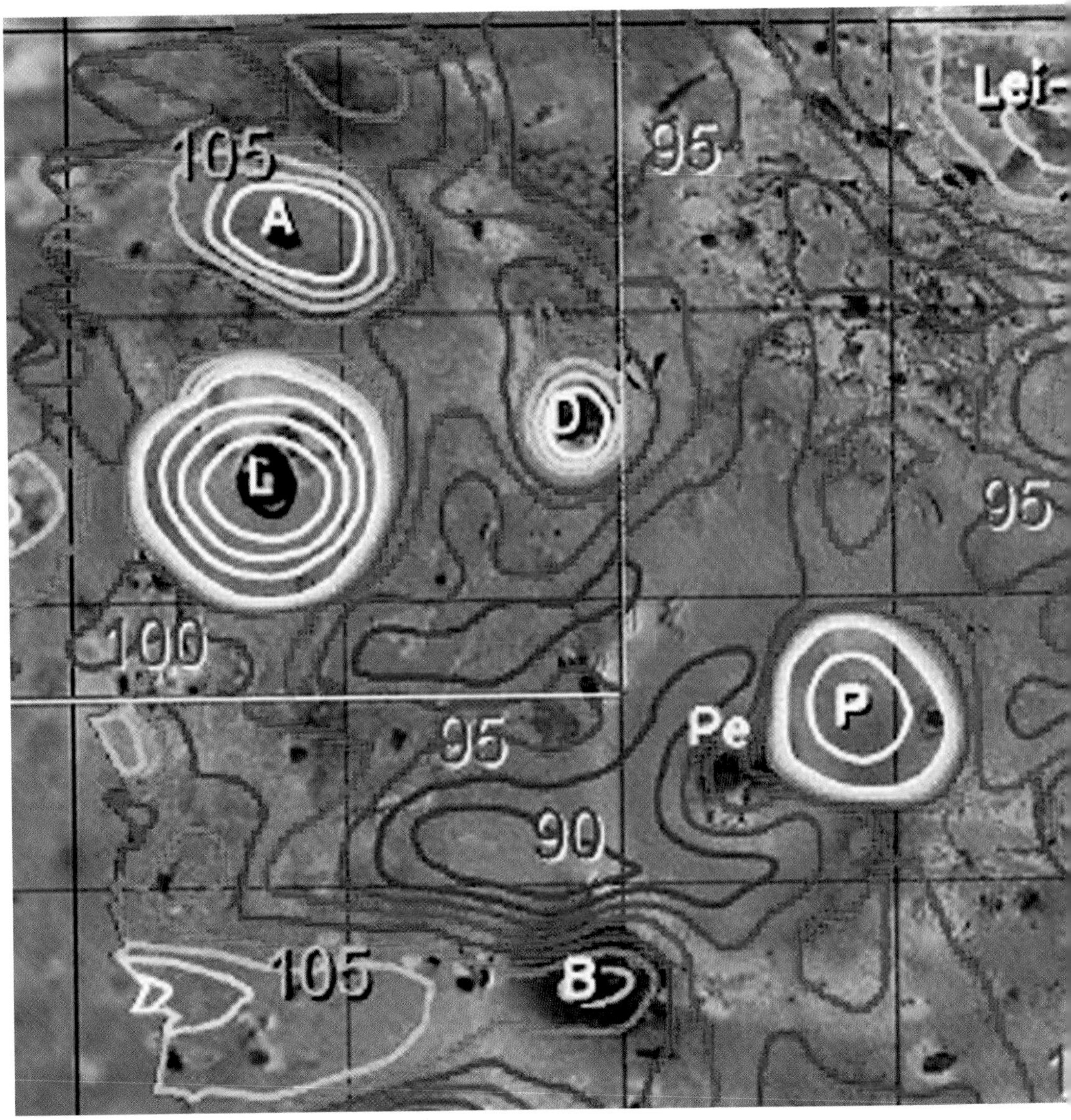

105
A
95
Lei-
D
L
95
100
P
Pe
95
90
105
B

IO In the decades since the Voyager flybys, the count of known active volcanoes on Io has grown from fewer than 20 to nearly 100. Many more lie dormant or have merely paused between eruptions. A temperature map of Io's night side (above) shows bull's-eyes over hotbeds of activity. Contours are labeled in the Kelvin scale: 90K is -297°F, and 105K is -270°F. The brightest are Loki (L), Amaterasu (A), Daedalus (D), Pillan (P), Pele (Pe), Marduk (M), Babbar (B), and a huge, old lava flow called Lei-Kung, which erupted sometime before the Voyagers arrived but is apparently still warm. The total amount of heat coming out of Io's interior is much larger than the global output of another highly volcanic world: Earth.

Fresh eruptions on Io are seldom hard to miss. Over a five-month period in 1997, a dark Arizona-size stain (opposite, upper) appeared around the vent of Pillan Patera. The deposit partly covered the ring of red sulfur about 800 miles wide surrounding an even-larger volcano called Pele, which itself changed in the interim. Note the fresh deposit (opposite, lower).

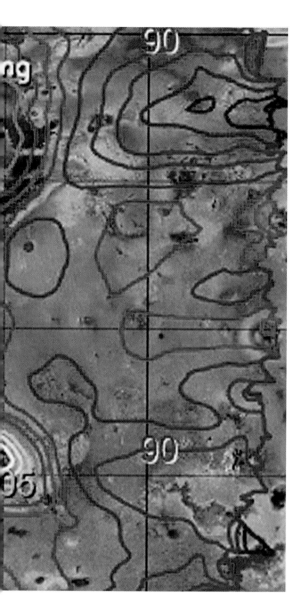

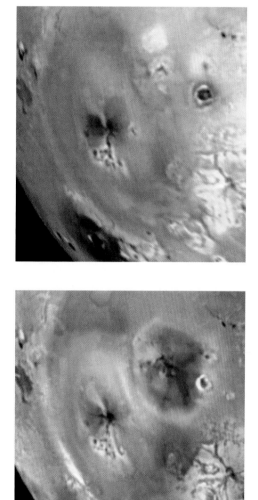

INTERPLANETARY
WANDERERS

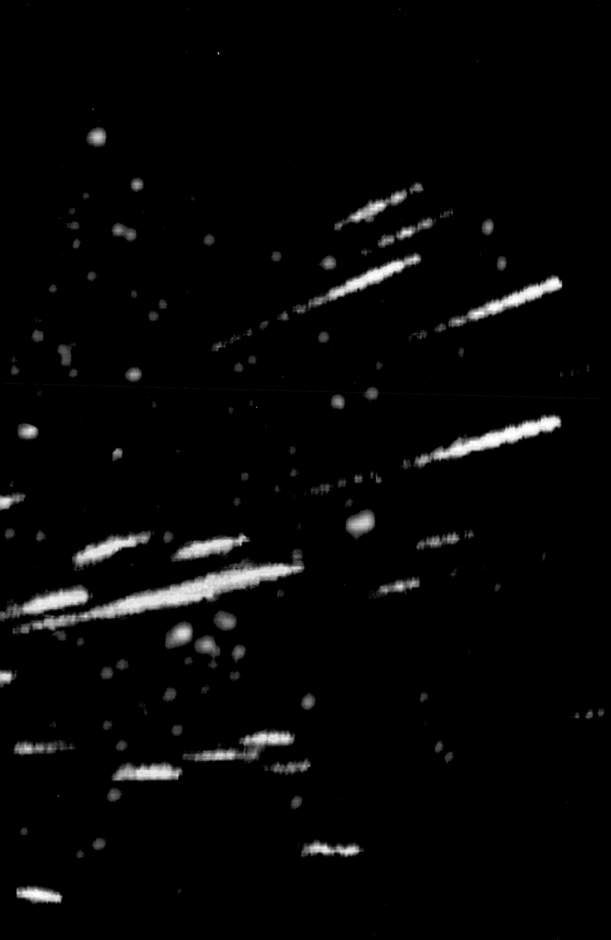

INTERPLANETARY
WANDERERS

The spring of 1997 was a special time for skywatchers everywhere when a city-size chunk of ice and rock from the distant fringes of interplanetary space came hurtling through the inner solar system. Comet Hale-Bopp, named for two backyard astronomers who discovered it, loomed large and bright in March's predawn twilight before swinging around to grace the evening skies of April and May. When they had first set eyes upon their namesake 21 months earlier, Alan Hale and Thomas Bopp each knew they'd found something special—but little did they realize how spectacular their find would become. It proved to be the brightest comet in a quarter century, a dramatic display of light and color. Some 70 percent of all U.S. adults glimpsed it, according to one survey. Before Hale-Bopp's arrival, however, most casual skywatchers had never seen a comet because few of these cosmic interlopers ever become obvious to the unaided eye. Yet dozens are spotted every year—many by dedicated backyard observers hoping to discover the next "big one" that will bear their name.

"COMET" IS DERIVED FROM A GREEK WORD FOR "LONG-HAIRED ONE," and many ancient peoples thought of them as flowing objects like "broom stars" or "stars with long feathers." Appearing unpredictably, these heavenly messengers were often considered omens of evil or danger. In the 4th century B.C., the Greek philosopher Aristotle proposed that comets were "fiery exhalations" high in Earth's atmosphere. However, from his careful observation of a comet in 1577, Tycho Brahe argued that comets inhabit interplanetary space. Edmond Halley, an English astronomer and mathematician, deduced that the bright comets seen in 1531, 1607, and 1682 were the same object. He then predicted that another visit would occur in late 1758—which it did, 16 years after his death. It's been known as Halley's comet ever since.

Today we realize that comets, together with asteroids, represent the leftovers of solar-system formation. Planet building was a messy, inefficient process, and only a fraction of the raw material available in the solar nebula ended up in something sizable. For example, the region between Mars and Jupiter probably started out with at least a hundred times more mass than remains there now as asteroids. Farther out, in the frigid, dimly lit fringes of the nebula, countless small bodies remained after the giant planets collected themselves. Most cometary bodies are far too distant and too small (the majority are only one or two miles across) to be seen with telescopes. But when they venture near the Sun they become transformed into diaphanous wonders.

PRECEDING PAGES: Comet Hale-Bopp dazzled with gas (blue) and dust tails in March 1997.
OPPOSITE: Time-lapse video records a pulse of Perseid meteors in August 1998.

COMET CLOUDS. Some comets, like Halley's, are termed periodic because their long, looping orbits around the Sun repeatedly carry them through the inner solar system. However, many others appear to come from exceedingly great distances, a trillion miles away or more. Throughout the 1800s astronomers believed that about a third of all comets somehow formed in interstellar space before being captured by the Sun's gravity. In the late 1940s a Dutch theorist, Jan Oort, calculated that these most distant bodies originate in an enormous swarm, or cloud, whose outer edge is more than a full light-year—6 trillion miles—away! He surmised that subtle gravitational forces, even the gentle tug of a star passing in our solar system's general vicinity, could perturb the motion of these distant comets and cause them to "fall" toward the Sun.

The appearances of bright comets have been recorded for thousands of years. This Babylonian clay tablet dating from 87 B.C. describes a visitor we now call Halley's comet, which passes near the Sun every 76 years.

Oort's comet cloud didn't start out so far away, however. In the early solar system, the outer planets' neighborhood teemed with icy planetesimals. Those that ventured too near Jupiter or Saturn were flung into interstellar space; others received gentler nudges. Some encountered Uranus or Neptune and drifted outward but didn't leave the solar system entirely, lingering at the limit of the Sun's gravitational influence. Seen from the outside, this vast reservoir would resemble a beehive in ultraslow motion, each comet taking millions of years to circle the Sun. From the swarm's outer edge the Sun is no more than a distant beacon; the temperature hovers near absolute zero. Astronomers can't detect the Oort cloud directly, but based upon the number of comets seen each year, it must hold trillions in cryogenic suspension.

About the time that Oort put forward his ideas, two other astronomers speculated that a second comet reservoir must exist. Kenneth Edgeworth (in 1949) and Gerard Kuiper (in 1951) independently theorized that orbital motion out beyond Pluto was too slow and the distances between objects too vast for them to collect into planets. Instead they must have remained as smallish bodies, billions of them, arrayed in a flattened disk far larger than the planetary realm. This has come to be called the Kuiper Belt—though some astronomers, in deference to the codiscoverer, use Edgeworth-Kuiper Belt. The first of its members was spotted in 1992, and since then telescopes have swept them up by the hundreds. A few Kuiper Belt objects are 400 to 500 miles across, rivaling the largest asteroids in size. A decade of searching has yet to find any farther away than about six billion miles. The Kuiper Belt, it seems, has a distinct outer edge.

Because comets at such distances have never experienced the Sun's warmth, they retain all the exotic compounds collected from the primordial solar nebula. Comets are thus Rosetta stones of the solar system's formation, and astronomers are eager to learn their composition. We can't yet visit the Oort Cloud or Kuiper Belt, but fortunately their representatives occasionally make the long trek to us after being nudged inward by gravitational shuffling.

OTHER WORLDS

We've come to think of a comet's solid core, or nucleus, as a "dirty snowball" thanks to keen insights made by Harvard astronomer Fred Whipple in 1950. Near the Sun, ices in the nucleus become warm and spew into space as a great cloud of dust-choked vapor, termed the coma, that hides the nucleus from view. Initially Whipple thought that the ices in comets were largely frozen water, and that's true. But some comets "turn on" when they get near the orbit of Jupiter, 400 to 500 million miles from the Sun, where the temperature is too cold for water to vaporize quickly. Cosmic chemists eventually identified other molecules that can escape from the nucleus readily even when water is still frozen rock solid: carbon monoxide (CO), which changes from ice to gas at -325°F, and the compound CN, sometimes called cyanogen, which must be a fragment of some other, larger "parent" molecule.

A depiction of Halley's comet, top center, appears in the celebrated Bayeux Tapestry, which commemorates the Norman conquest of England in A.D. 1066.

The most scientifically prized comets are first-time or infrequent visitors to the inner solar system. These "fresh" comets can spew vast quantities of gas into space. The pressure of sunlight, though very slight, is enough to push the gas and dust away, forming a tail. Sometimes a comet will have two tails: a pearly white one containing dust, and a second bluish one with gas molecules that have become charged, or ionized, by sunlight. The ionized gas tail always points directly away from the Sun (to within a few degrees), whereas the dust tail can be angled as much as 45° with respect to the Sun-comet line.

A comet sometimes leaves a trail of clues to its true identity. As it rushes along, dust and larger particles shed by the nucleus eventually spread out along its orbit. And if we run into that ribbon of debris somewhere along Earth's orbit, the comet's castoffs dazzle us with streaks of meteor light high in our atmosphere. The most well-known of these meteor showers arrives like clockwork every August. It's known as the Perseid shower, because the meteors all seem to radiate from a point in the constellation Perseus. Those flashes mark the blazing end to dust grains shed by a periodic comet named Swift-Tuttle.

To Catch a Comet. Studying meteors and comets from afar can tell astronomers only so much, however. What they've really wanted all along is a good look at a comet's icy heart—and the only way to do so was to send spacecraft to one. So when Halley's comet raced through the inner solar system in 1986, a group of the world's space powers dispatched a small armada to greet it. Lack of funds kept the United States from joining in. But Japan launched two craft, named Suisei and Sakigake, which monitored Halley from afar. Others took a more direct approach: A European craft called Giotto and two from the Soviet Union named Vega 1 and Vega 2 headed straight into Halley's dusty coma.

Giotto drew the most formidable assignment, skirting past the seething, sputtering nucleus at 43 miles per second. At that speed a grain of sand can pierce a wall of aluminum three inches thick. Miraculously, Giotto survived, and it taught us much about Halley's legendary comet. For one thing, at its heart is a mountain ten miles long and half as wide, several times larger than the average comet nucleus. And far from being covered with ice, the heart of Halley was truly black, darker than asphalt or coal. Giotto's camera recorded several geysers of gas and dust, the source of the dense coma and long tail. More than 30 different chemical compounds were escaping from the icy core, and about a third of the dust particles consisted almost entirely of carbon, hydrogen, oxygen, and nitrogen—key ingredients for virtually all organic matter.

In many cultures, ancient and modern, comets represented portents of evil or doom. In this 16th-century woodcut, the Great Comet of 1528 is trailed by an assortment of daggers and swords.

A quantum leap in our understanding of comets should come in the next decade, as four spacecraft make close-up studies. In January 2004, NASA's Stardust spacecraft will dash through the coma of Comet Wild 2 at nearly four miles per second, sweeping up the fragile dust grains in a spongelike glass "foam" called aerogel. With luck, the spacecraft will survive its mad dash and return the coma-infused collector to Earth two years later.

Meanwhile, a different kind of sampling will be attempted in July 2005, when a second NASA craft, named Deep Impact, sends a 770-pound copper bullet crashing into the nucleus of comet P/Tempel 1 at more than six miles per second. The scientific plan is simple: Watch to see what happens next. The carrier ship will monitor the collision and analyze the debris that sprays out from it. A third NASA mission, Contour, will cruise around the inner solar system between 2003 and 2006—long enough to sweep within 60 miles of two different active comets.

European scientists have designed a mission called Rosetta to fly alongside and around Comet Wirtanen for nearly two years beginning in November 2011, mapping the icy nucleus and monitoring its response to intense heat and light as it nears the Sun. Swooping to within about a half mile of the comet's heart, Rosetta will release a small instrumented lander that will dig into the nucleus and assay the compounds that give comets their diaphanous character and beauty.

VERMIN OF THE SKY. Because asteroids were unknown until the dawn of the 19th century, they do not share comets' long, rich history. They were never seen by ancient skywatchers as portents of doom. No Greek philosopher ever pondered their cosmic significance. Nonetheless, the discovery of the first asteroids is a curious, colorful tale involving a Hungarian nobleman, an Italian monk, and the self-styled "Celestial Police." Born in 1754, Baron Franz Xaver von Zach was a budding astronomer with a keen interest in the Titius-Bode "law," a simple arithmetic formula that predicted the existence of a planet between Mars and Jupiter. In 1800 Zach organized fellow astronomers, who called themselves the Celestial Police, to track down the missing orb.

Meanwhile, the Theatine monk Giuseppe Piazzi had been compiling a new star catalog from his observatory in Palermo, Italy. On January 1, 1801, Piazzi found a conspicuous object in what later proved to be in the "missing" planet's orbit. He named his discovery Ceres, for the patron goddess of Sicily, and astronomers rejoiced at the apparent validation of Titius and Bode. However, Ceres was hardly alone: By 1807 three more sizable bodies had turned up between Mars and Jupiter. The count continued to rise, passing a hundred in 1868 and a thousand in 1921.

Too small to show any detail when viewed telescopically, minor planets were once considered "vermin of the sky" because they often left trailed streaks among the pinpoint stars of celestial photographs. But in time astronomers learned to interpret the gleam of asteroidal light. Their subtle differences in color and the fact that their surfaces preferentially reflect certain wavelengths of sunlight offer compositional insights.

More clues came from meteorites, which are in effect free samples of various asteroidal bodies. Untold thousands of meteorites land on Earth each year, though very few are ever found and even fewer are seen to fall. Although the largest and most spectacular specimens in museum collections are seared hunks of nearly pure iron-nickel metal, about 80 percent are stony-looking objects called chondrites. "Chondrite," derived from a Greek word for "grain," alludes to the tiny round chondrules of once molten silicate often found in abundance within them. These stones—and especially the rare carbonaceous chondrites—are primitive survivors from the earliest times of solar-system history. Other meteorite classes, like the irons and stony-irons, come from larger precursors that were once partially or completely molten interiors. But few asteroids remained wholly intact for four and a half billion years: At some past time virtually all of them must have collided with one or more of their siblings so violently that shattered fragments are now strewn throughout interplanetary space.

Taken together, these lines of evidence tell us that there are different compositional niches within the asteroid belt. Within two and a half astronomical units (230 million miles) of the Sun, some asteroids look to be dry mixtures of silicate minerals, things that would pass for "rocks" on Earth; others show only weak hints of silicates, apparent matches to the iron meteorites. Farther out, most asteroids are darker, as if infused with carbon, and show signs of water (clay minerals). This compositional variety jibes well with theoretical scenarios of what took place within the asteroidal zone in primeval times. The condensing solar nebula created a multitude of primordial planetoids outside the orbit of Mars. Those closer in garnered little water but plenty of heavy elements—including several short-lived isotopes. The largest masses grew hot enough from radioactive decay to melt completely, stratifying as they cooled into metallic cores and silicate mantles. The most distant objects remained cold and primitive, and they closely resemble the composition of the solar nebula itself.

Left to themselves, all these minor planets might have gathered into a single major planet. But Jupiter and its powerful gravity intervened, turning what had been orderly, circular orbits into an overlapping jumble. Within several million years more than 99 percent of the asteroidal precursors had been drawn into the giant planet, cast into the Sun, or ejected from the solar system altogether. Jupiter's runaway growth precluded the rise of any serious planetary contenders between its orbit and that of Mars, and today all that remains there are scattered remnants that attest to the giant planet's gravitational primacy.

Two centuries after their discovery, we know the original "big four" asteroids—Ceres, Pallas, Juno, and Vesta—share their interplanetary niche with at least 100,000 siblings. With this growth in numbers has come a better understanding of what asteroids are really like, thanks to state-of-the-art observing tools like powerful radar systems and the Hubble Space Telescope. Astronomers have learned that some asteroids spin in just a few minutes, while others take days. Many are elongated in shape. A select few even have small satellites.

But there's a limit to what can be learned from dissecting asteroidal light, and it was only a matter of time before spacecraft would be dispatched to satisfy our scientific curiosity. During the early 1990s the Jupiter-bound Galileo orbiter flew past medium-size objects named Gaspra and Ida. Their grayish surfaces, thoroughly pocked with craters, have been exposed to space for at least hundreds of millions of years. They appear to be fragments of much larger bodies that broke apart well along in solar-system history. One unexpected bonus was spotting a small moonlet, Dactyl, orbiting close to Ida. A later mission, known by the burdensome name of Near Earth Asteroid Rendezvous (NEAR), swept past dark, carbon-rich Mathilde en route to the primary target, Eros, which it orbited at close range for an entire year.

Not all asteroids are found in the asteroid belt. Many thousands, called the Trojans, are trapped astride Jupiter's orbit in gravitational Bermuda Triangles, one well ahead of the planet and one behind it. A handful leak past Mars and into the inner solar system. Others drift out among the outer planets, where some otherwise inert objects have the kind of elongated, tilted orbits typical of comets. The "asteroid" Chiron, found in 1977 drifting in a 51-year-long orbit beyond Saturn, began sporting a coma in the late 1980s. Clearly a huge comet, it has a diameter of 120 miles. Likewise having a cometlike trait, an asteroid called Phaethon shares the same orbit as the Geminid meteor stream.

"Scientists have a strong urge to place Mother Nature's objects into neat boxes," observes Donald K. Yeomans, an orbital specialist at the Jet Propulsion Laboratory. "Within the past few years, however, Mother Nature has kicked over the boxes entirely, spilling the contents and demanding that scientists recognize crossover objects—asteroids that behave like comets, and comets that behave like asteroids. As a result, the line between comets and asteroids is no longer clearly drawn." But the confusion is welcome, for once we understand the origin and evolution of these interplanetary wanderers, the solar system's history will become an open book for our eager reading.

OPPOSITE: The cloud-like coma of Comet Hale-Bopp (color-coded based on its brightness) in 1996.

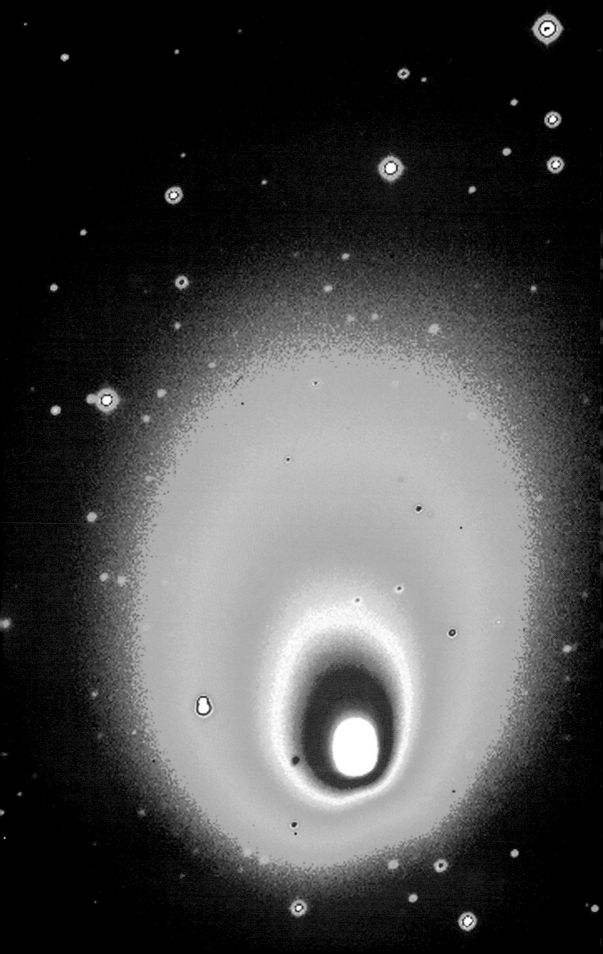

The Great Crash of 1994

One of the greatest events in the history of astronomy began with some bad film, a cloudy night, and four frustrated observers.

On Palomar Mountain in southern California, virtually in the shadow of the legendary 200-inch Hale telescope, stands a small research telescope used almost exclusively for hunting comets and asteroids.

The prospects for discovery seemed bright when the observing team of Eugene Shoemaker, his wife Carolyn, and long-time collaborator David Levy, along with visiting French astronomer Philippe Bendjoya, arrived on March 21, 1993, to use the telescope. March 22 was their first night of viewing, but they discovered that their film had become fogged by accidental exposure to light. On the following night, March 23, they optimistically began their second night of viewing with good film.

Their optimism soured, however, soon after cracking the dome open. Thin clouds overhead heralded advancing overcast. Resigned to the fact that any pictures they took would be compromised by the clouds, they decided not to waste any of their good

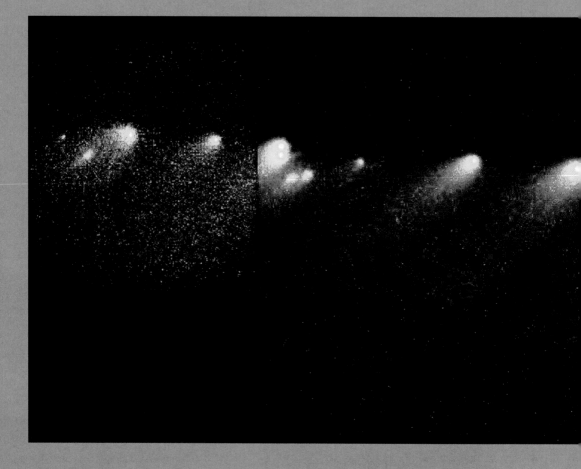

film. Instead, using some of the film that was slightly fogged, they recorded one more area of sky two times. Then the clouds thickened, and they called it a night.

Two days later, after the film was developed in the basement darkroom, Carolyn examined the negatives closely. Off to one side, not far from Jupiter, was a fuzzy streak with a series of tails and thin lines at either end. "I don't know *what* this is," she said, bolting upright. "It looks like…like a squashed comet."

More powerful telescopes later revealed the blur to be a line of little comets arrayed like pearls on a string, each sporting its own tail. Based on other observations in the weeks thereafter, astronomers concluded that a single, giant iceball had strayed to within 14,000 miles of Jupiter on July 8, 1992, and had been torn apart by the planet's gravity.

But even more stunning was the revelation that the shattered remains of Comet Shoemaker-Levy 9, or S-L 9, as it became known, were destined to strike Jupiter itself in July 1994.

A cruel twist of geometric fate placed the target zone on Jupiter's far side, just out

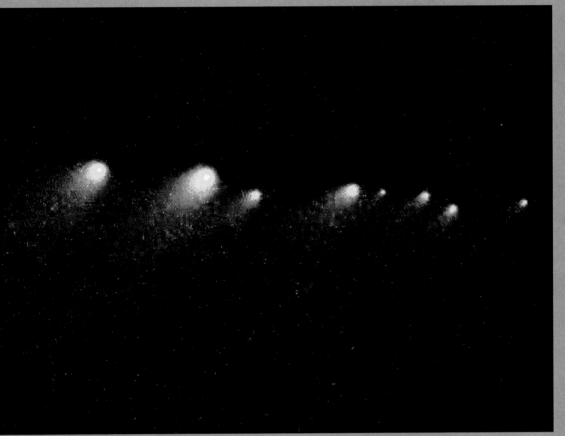

All strung out: Comet Shoemaker-Levy 9 in January 1994.

of sight from Earth. Computer-aided simulations tried to anticipate what would happen during each high-speed splash into Jupiter's atmosphere, and what might be seen once the impact sites rotated into view. Some modelers suspected that the fragments were large, at least a mile across, and would strike with the kinetic-energy equivalent of a hundred billion tons of TNT—or more.

Others countered that Jupiter might well swallow the icy shards without a trace. But no one could predict the outcome with certainty. Never before had such an event been witnessed, and never before had so many of the world's telescopes—from mountain-top behemoths to humble backyard tubes on rickety tripods—turned their gaze to the same spot of sky. Even the orbiting Hubble Space Telescope and Jupiter-bound Galileo spacecraft were commandeered for the comet's death watch. Thanks to its location in interplanetary space, Galileo was able to view the target zone directly when the impacts commenced

Astronomers don't always fare well during highly publicized celestial events, so caution generally pervaded the pre-crash pronouncements. However, for once the celestial fireworks were nothing short of spectacular. Over a period of six days, beginning July 16th, a score of cometary fragments bombarded Jupiter at 40 miles per second. Several created fireballs roughly 2,000 miles high, tall enough to peek around the planet's limb and be spotted by the Hubble Space Telescope.

Observers sat open-mouthed as tremendous fireballs rotated into view and blazed into their telescopes' infrared-sensitive detectors. The superheated gas in the target zones had temperatures approaching 20,000°F or more. As the conflagrations cooled, most left huge dark stains in the Jovian atmosphere. In some cases these splashes of sooty debris were larger than the entire Earth.

"I feel sorry for Jupiter," quipped astronomer Heidi Hammel. "It's really getting pummeled."

The black scars smeared out and faded after several months, though researchers continued to analyze the impact's consequences for years thereafter.

Spectroscopists dissected the light from the superheated blast plumes in the hope of deducing something about the composition of the planet's atmosphere. But it proved difficult to disentangle which compounds had come in with the kamikaze fragments and which belonged to Jupiter.

"Trying to learn about comets from this impact is like trying to learn entomology from the bugs on your car's window," observed NASA astronomer Paul Weissman.

Much effort went into pinning down the original size of Shoemaker-Levy 9 and of the pieces it became. No consensus was reached, though all the havoc wrought on Jupiter—the towering, incandescent plumes and the globe-girding tangle of dark atmospheric bruises—could easily have been the handiwork of an errant iceball no larger than one mile across.

An object of this size probably strikes Earth every 100,000 years on average, and, judging from the consequences seen on Jupiter, it would not be an event awaited with eager anticipation.

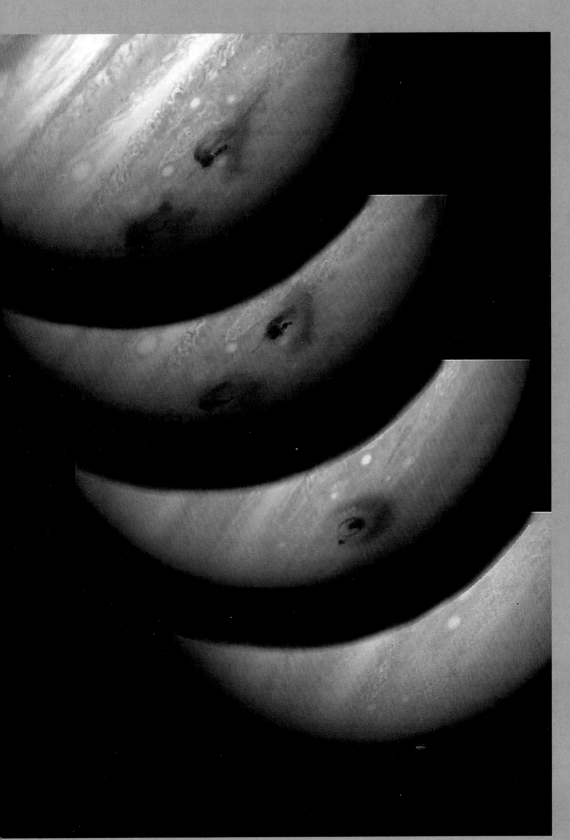

These images, taken over five days, record Jupiter's response to being hit.

FOLLOWING PAGES: Comet Hale-Bopp presages the dawn
over Italy's Val Parola Pass in March 1997.

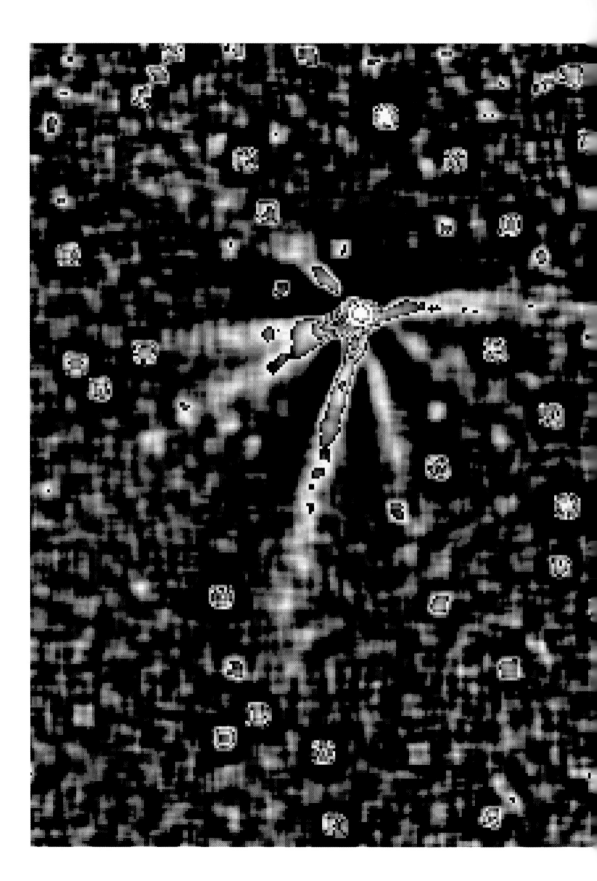

OTHER WORLDS

COMETS When a comet nears the Sun, its frozen ices vaporize and escape to space however they can: as soft gauzy clouds or as tight streams. Exaggerated coloring (left) reveals at least seven gas-powered jets that emanated from the nucleus of Comet Hale-Bopp in late 1996, nearly half a year before becoming its brightest. (Other spots are faint stars and artifacts.)

Many small comets never appear in the night sky but still pass near—or into—our daytime star. These kamikaze comets, known as Kreutz Sungrazers, are believed to originate from a single large parent comet that broke apart near the Sun perhaps 2,000 years ago. Astronomers witness the fragments returning to a fiery fate (above) by using a specially masked camera aboard the SOHO spacecraft (white ring denotes Sun's location). Besides Halley's comet (right), to date no comets have been photographed at close range.

Meet Mr. Comet

Despite being in his mid-90s, Fred Whipple can be found most weekdays in his sunny corner office at the famed Smithsonian Astrophysical Observatory in Cambridge, Massachusetts. He stopped riding his bicycle to work a few years ago, but age has not dulled his keen interest in space exploration and research.

Whipple and the SAO have been inseparable since 1955, when its offices were moved up to Cambridge from Washington and he became its director. By then, Whipple had already achieved fame for his "dirty snowball" theory, a radical new way of looking at comets. For nearly a century these cosmic visitors had been imagined as flying sand banks, because their graceful tails were known to shed a multitude of dust particles that produced meteor showers in Earth's atmosphere. But no one could explain all the fluorescing gas seen in comets' tails, a problem that gnawed at Whipple throughout the 1940s.

Intent on finding the answer, he soon concluded that comets must somehow store lots of gas inside their nuclei. "Ice was the obvious answer, and that led me to the dirty-snowball model," Whipple recalls. "If you put dust into a deep freeze of ice, particularly ices composed of water and possibly carbon dioxide or ammonia, then you have a supply that could last for a huge period of time. That seemed pretty obvious to me."

Whipple found more evidence for his nascent idea from Comet Encke, which circles the Sun every 3.3 years. Astronomers had known for some time that this comet's orbit was slowly shrinking but could not explain why. In contrast, Halley's comet came back late in 1910. Whipple, at last, had the answer: "When the Sun shines on an icy mass in space—in a vacuum—the gas comes off at about half a kilometer [0.3 mile] per second. That's a lot of force acting on the nucleus." The escaping gas was acting like a gentle rocket, nudging the nucleus into an ever-changing orbit. Knowing the rate of the orbit's change and the

amount of force exerted by the gas, he deduced that Encke's icy nucleus was no more than one to two miles across.

Whipple didn't refer to comets as "dirty snowballs" when his conclusions were published in 1950, but the nickname—and the idea—stuck. In March 1986 he was on hand in Moscow and, a week later, in Germany, when Soviet and European spacecraft swept past Halley's comet and revealed its nucleus to be a geyser-spouting mass of dark rock and ice. Now this elder scientific statesman is on the research team for NASA's Contour mission, which will fly through the coma of Comet Encke in 2003 and Comet Schwassmann-Wachmann 3 in 2006. Always curious, Whipple hopes to be around for those events, too.

Comet expert Fred Whipple teaches a class.

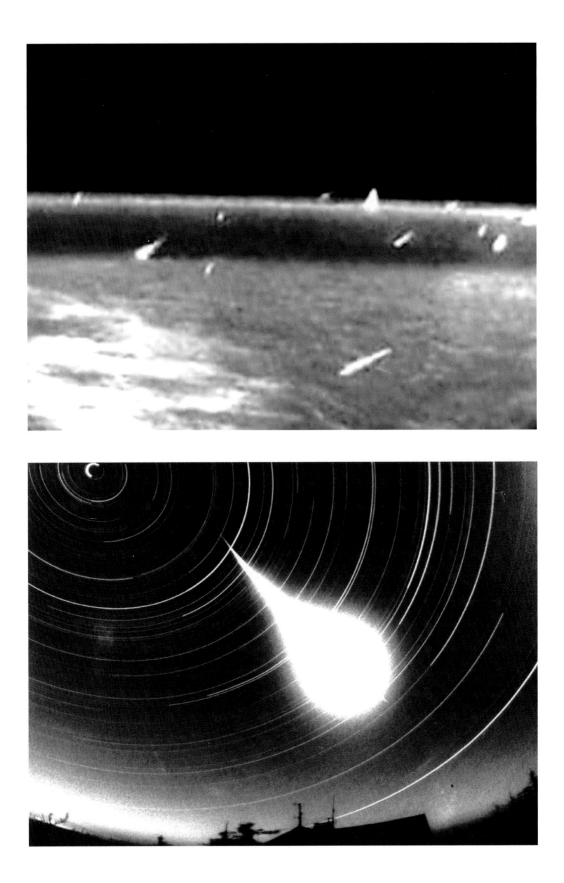

METEORS

"Shooting stars" are flecks of comet or asteroid
that have strayed onto a collision course with
Earth. Sometimes these cast-offs form a stream
along a comet's orbit, cascading into our atmosphere as concentrated meteor showers whenever
Earth crosses their path. A dazzling display of the
Leonid meteors (opposite, upper), shed long ago
by Comet Tempel-Tuttle, was captured from an
orbiting spacecraft in November 1997. An even
bigger "storm" followed in 1999.

Most meteors are pea-size or smaller, but they can
be much bigger. Stone-size arrivals create dazzling
fireballs (opposite, lower). Anything smaller than an
oil drum is unlikely to reach the ground. But even
bigger strikes do occur, if only rarely. Some 210
million years ago, a small asteroid (or large comet)
slammed into northern Canada and created the
60-mile-wide Manicouagan crater (below).

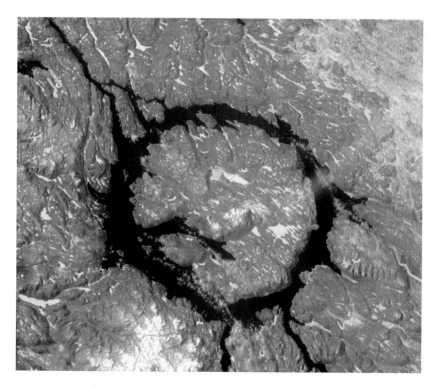

FOLLOWING PAGES: Leonids are known for glowing trains that persist long after the
meteors' flash. A long-lived Leonid "glowworm" is probed by a laser experiment from
an observatory in 1998.

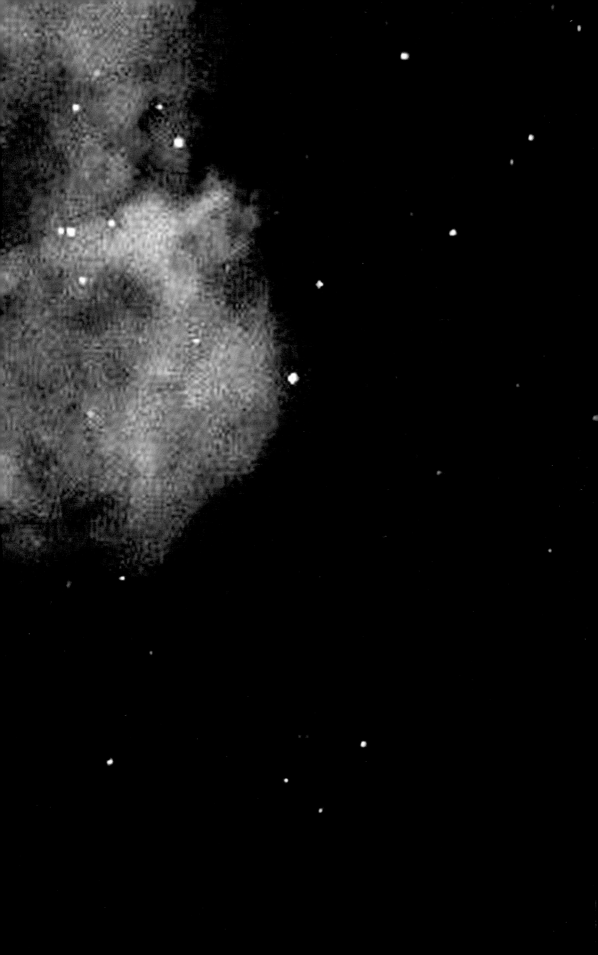

Heaven on Earth

The morning of January 18, 2000, dawned cold and clear over Canada's remote Yukon Territory. But at 8:43 a.m. winter's calm was shattered by a dazzling fireball that outshone the Sun as it streaked southward across the sky.

Then angry detonations shook the ground, signaling that the intruder had exploded in midair. Police switchboards in Whitehorse and the surrounding towns lit up too, as local residents—many clearly worried—called with various details of the startling spectacle. Some had heard sizzling sounds, while others reported noxious, sulfur-laced smells.

When news of the fireball reached Peter G. Brown, a meteor specialist at the University of Western Ontario, he knew there could be more to this event than celestial fireworks. Such airbursts often end with meteoritic fragments falling to the ground below, and this one was powerful enough to have registered in the sensors of orbiting defense satellites.

So residents in the area were alerted to keep an eye out for dark, suspicious-looking rocks. Brown was cautiously hopeful of something turning up: after all, at that time of year much of the region is frozen solid—and white.

One week later, Jim Brook was bounding home in his truck across the frozen surface of Taku Arm, a spur of Tagish Lake just south of the Yukon-British Columbia border. An amateur geologist and local fishing-camp operator, Brook had heard all about the meteor, and he figured any rock sitting atop the lake would have fallen there from above. Sure enough, he spotted some despite the fading daylight.

"I was watching closely for meteorites and suspected their identity as soon as I saw them, although I'd been fooled several times by wolf droppings," Brook recalls. "It was obvious what they were as soon as I picked one up."

He returned the next day, collecting several dozen fragments with a total weight of about two pounds. Several days later a storm moved in, burying the undiscovered pieces under deep snow.

Brook didn't realize it at the time, but he had become a pivotal player in one of the most important meteorite finds in decades. The Tagish Lake stones are examples of a rare meteorite type known as carbonaceous chondrites: crumbly, charcoal-black masses infused with carbon, water, and a host of primitive organic compounds dating to the earliest era of solar-system history.

But carbonaceous chondrites are rarely found, because they break up easily during their atmospheric passage or quickly erode once on the ground. That explains why they represent only 3 percent of known meteorites—and why Brook's discoveries have been celebrated by cosmic chemists worldwide.

As January's cold gave way to spring's thaw, a field team led by geologist Alan Hildebrand from the University of Calgary and Brown joined Brook for a more thorough search of the still frozen lake. By then the snow was gone, but many of the black shards had become encased in ice.

Toiling for nearly three weeks, sometimes using chain saws to hack out huge meteorite-bearing popsicles, the team recovered another 410 pieces from the rapidly thinning ice cover. Taku Arm's waters have now

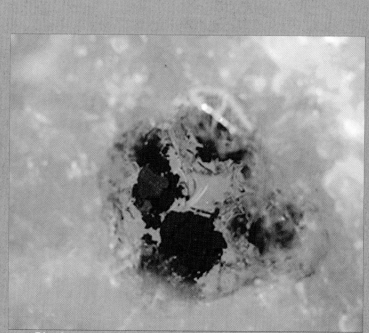
Black and crumbly, a meteorite lies encased in lake ice.

swallowed any remaining fragments, and only a handful have been found in the rugged, heavily forested hills surrounding the lake.

Based on the field work, the pictures and videos from eyewitnesses, and the satellite data, Brown and Hildebrand have reconstructed the events of that remarkable January day.

Arriving from a point roughly midway between Mars and Jupiter, a chunk of primitive asteroid 15 to 20 feet across and weighing up to 200 tons slammed into Earth's atmosphere at 10 miles per second, delivering the kinetic-energy equivalent of 5,000 tons of TNT. It exploded at an altitude of 20 miles, scattering perhaps 10,000 fragments along its atmospheric trajectory in a narrow ellipse 3 miles wide and at least 10 miles long.

Meanwhile, the stones themselves are revealing big surprises—due, in no small part, to Brook's textbook collection technique. (He never touched the stones by hand and even kept them frozen until they could be shipped to specialists.)

The Tagish Lake meteorites represent a particularly rare subtype of carbonaceous chondrite, one that is perhaps even compositionally unique, and it ranks as one of the oldest and most primitive chunks of solar-system matter ever studied. Some of its carbon is in the form of organic matter, and there's even a smattering of interstellar diamond dust—all virtually certain to be free of terrestrial contamination.

Geochemists have not gotten their hands on such a prized specimen in decades (two previous falls of this type occurred in 1969), and they're sure to use every trick in their analytical arsenal to unlock its many secrets.

"The more pristine meteoritic material of this type we can analyze," notes Jeffrey N. Grossman of the U.S. Geological Survey, "the better we will understand how our solar system formed and how the remnants of others came to be incorporated into it."

Indeed, the Tagish Lake meteorite find is tantamount to a cosmic Rosetta stone that fell to Earth and into meteoriticists' waiting hands. Brown, who still can't believe all this good fortune, exults, "It's the find of a lifetime."

OTHER WORLDS

ASTEROIDS

Astronomers calculate that Earth's orbit is crossed by nearly a thousand asteroids at least two-thirds of a mile across. Objects of this magnitude would strike Earth with the kinetic-energy equivalent of 100 billion tons of TNT. One of the asteroids in this "potentially hazardous" category is named Toutatis. About three miles long, Toutatis (shown above in the foreground of a computer simulation of it in relation to Earth) will pass within a million miles of Earth in September 2004, coming closer than any known asteroid over the next 30 years. Inevitably, Earth will once again be hit by an asteroid or comet large enough to cause mass extinctions, as probably happened 65 million years ago. But smaller collisions with Earth occur much more frequently. Kiloton-class meteoroids explode in our atmosphere several times each year. An object the size of a large warehouse, delivering the force of a 50-megaton bomb, probably arrives every few centuries on average. Sometime within the next million years we'll endure a hit powerful enough to gouge out a crater the size of 14-mile-wide Gosses Bluff in Australia (left).

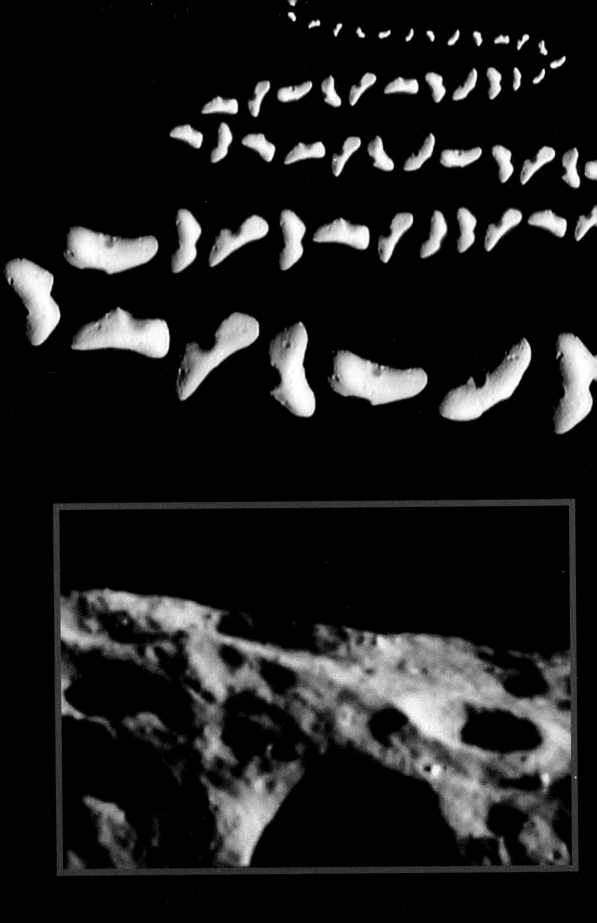

OTHER WORLDS

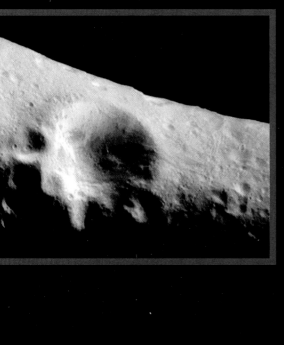

E R O S Know your enemy. A NASA
spacecraft called the Near Earth Asteroid
Rendezvous, or NEAR, spent a year orbiting
an asteroid called Eros. Although it poses no
near-term threat to us, Eros is similar to the
kind of asteroid typically found crossing
Earth's orbit. During the early stages of
NEAR's approach (upper left), Eros appeared
as a small blob, but surface details came into
sharper focus as the spacecraft closed in.
Shaped like a ballerina's slipper (above), Eros
is 20 miles long and about 8 wide. One prominent feature is a 3-mile-wide crater named
Psyche (left). Mission scientists still do not
understand why the asteroid's surface is
littered with thousands of boulders (far left,
bottom). NEAR-Shoemaker (renamed to honor
asteroid specialist Eugene Shoemaker)
eventually landed on the asteroid's surface

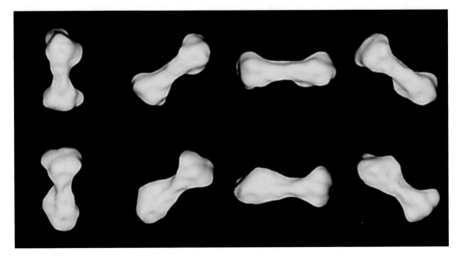

OTHER WORLDS

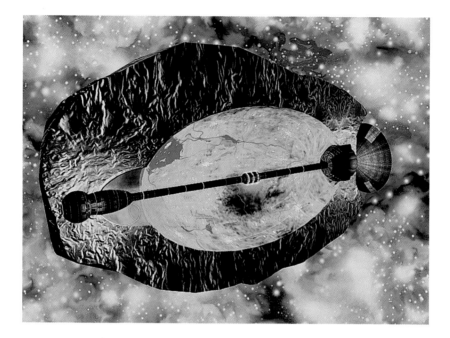

MISSHAPEN The first asteroid was discovered more than 200 years ago, but the true nature of these rocky objects remained a mystery until the space age. While en route to Jupiter, the Galileo spacecraft found that 35-mile-long Ida (left, upper) has a small satellite, now named Dactyl. Probably a mixture of rock and metal, Ida's interior is quite porous—making it an unlikely habitat for future space travelers (artist's rendition above). More unusual is asteroid Kleopatra (lower left), 135 miles long, whose dog-bone shape suggests that it is a leftover from an ancient, violent collision. One of several dozen asteroids whose coloring indicates metal content, Kleopatra was probed while 100 million miles away by powerful ground-based radar telescopes. Astronomers now realize that irregular, elongated shapes like this are rather common among asteroids.

CLOSE CALL Nearly two and an half miles across, the huge rock known as 1999 JM_8 silently passed only 5.3 million miles from Earth in early August 1999, even though it was completely unknown before the preceding May. As it neared Earth, astronomers bounced radar pulses off the object, and the resulting echo-based images (below) showed it to be a formidable-looking body.

"The discovery of this object weeks before its closest approach was a stroke of luck," notes astronomer Lance Benner. "The asteroid won't come this close again for more than a thousand years." The density of its craters suggests that the surface is rather old, geologically speaking, and not just a chunk broken off a bigger asteroid. For many decades astronomers did not understand how asteroids like 1999 JM_8 made their way to Earth—were they propelled our way by a distant explosion, as depicted on a 1930 magazine cover (opposite)? The real reason is much less dramatic: We now realize that there are several "escape hatches" in the asteroid belt, locations where the gravitational pull of Jupiter can redirect objects into long, oval-shaped orbits. Once they leave the belt's relative safety, they are destined to crash into a planet, dive into the Sun, or be ejected from the solar system altogether.

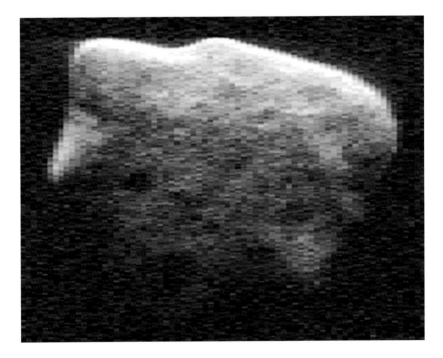

Science WONDER Stories

HUGO GERNSBACK Editor

March

25 CENTS
Canada 30¢

"Before the Asteroids"
by HARL VINCENT   GAWAIN EDWARDS   HAROLD S. SYKES   FRANK BRUECKEL

THE RED PLANET

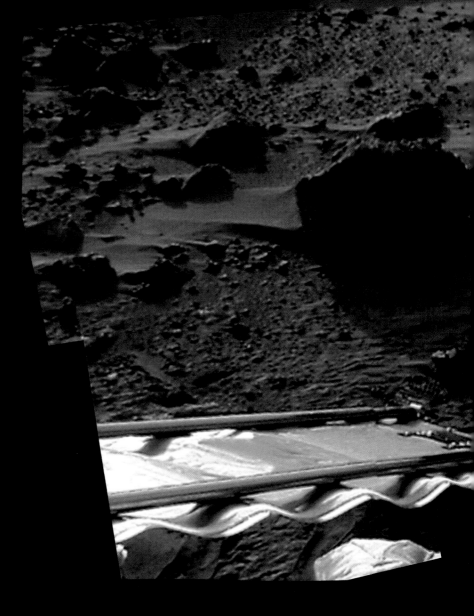

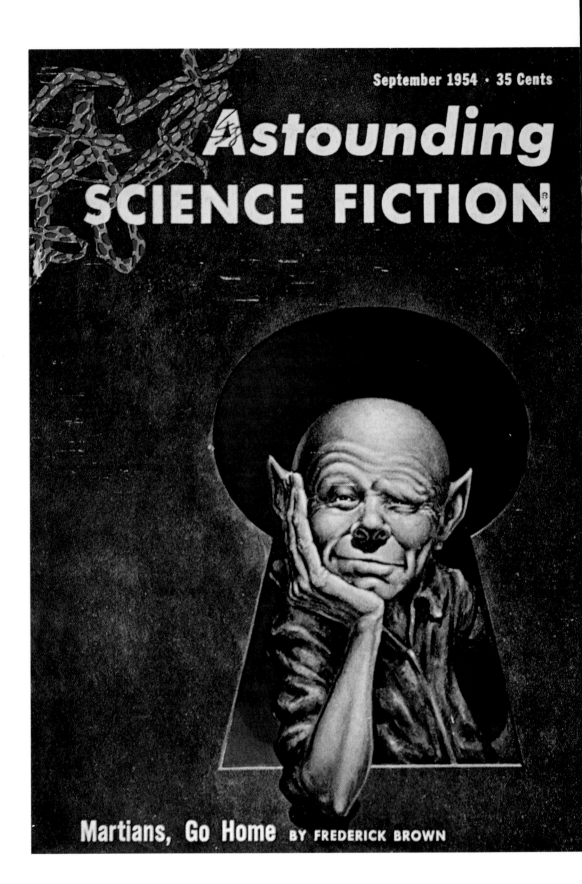

September 1954 · 35 Cents

Astounding
SCIENCE FICTION

Martians, Go Home BY FREDERICK BROWN

THE RED PLANET

In Ray Bradbury's classic *The Martian Chronicles,* June 2001 is the date on which the "Fourth Expedition" of astronauts arrives on Mars and finally establishes a human presence on the planet. Gaining that foothold took longer than expected because the first three crews mysteriously vanished without a trace (the Martians had killed them all). Undaunted, space officials back on Earth kept dispatching spaceship after spaceship until success was achieved.

Our real-life efforts to explore Mars have followed much the same plot. We've always looked toward the red planet with fascination, and as soon as technology allowed us to send robotic emissaries there for a closer look, we did. But the road to Mars was rocky. From the first attempts in 1960 until the triumphant Viking landings in 1976, most probes fell short of their goals because of exploding rockets, mechanical failures, missed orbits, and inexplicable disappearances. The success rate since the Vikings has been little better: the Soviet Union/Russia and NASA each lost three more spacecraft during the late 1980s and 1990s. As the 20th century closed, the score stood: Earthlings 12, Mars 25.

Nonetheless, among those dozen "wins" were some spectacular successes. In 1971 Mariner 9 slipped into orbit around Mars and provided more than 7,300 highly detailed images of the entire globe. The Vikings' twin orbiters showed us the planet's geologic variety from high above, and its twin landers surveyed the landscape from ground level and tested the surface for signs of life. Mars Pathfinder and its small rover, Sojourner, had a worldwide audience when they bounded onto an ancient floodplain in July 1997. Several weeks later, Mars Global Surveyor took up residence in orbit and ultimately taught us more about the red planet than in some respects we know about Earth.

Imagine for a moment that you have set off to inspect Mars as future astronauts might: from an observing station on Deimos, one of the planet's two small moons. Discovered in 1877, Phobos (Greek for "fear") and Deimos ("terror") are tiny by satellite standards, 16 and 10 miles across, respectively. Their dark, cratered surfaces look remarkably similar to those of asteroids—and perhaps they are asteroids, captured by Mars early in solar-system history. Phobos zips around its orbit in just 7⅔ hours, whereas more distant Deimos takes four times longer. These periods, compared with the 24⅔-hour spin rate of Mars itself, means that the two bodies appear to move across the sky in opposite directions as seen from the Martian surface.

Comfortably perched on Deimos, you glide over the planet's equator at an altitude of about 12,000 miles—perfect for taking in the entire globe in one

PRECEDING PAGES: Sojourner, an automated rover, sets out to explore the Martian surface.

OPPOSITE: Little green men from Mars, like this one, have long been a staple of science fiction.

sweeping spectacle. You are immediately struck by several things. First, the horizon looks distinctly round, because Mars's 4,220-mile diameter is only about half that of Earth. You then notice that details on the ground below appear sharp and distinct because the planet's carbon-dioxide atmosphere is very thin, equivalent to a 112,000-foot-high altitude over Earth. The surface is divided into large tracts of varying brightness, some light hued and others dark, continent-size colorings that can sometimes be glimpsed from home through a backyard telescope. And the ground really does look red, or at least reddish brown, thanks to an abundance of iron oxides—rust—in the Martian soil. Finally, your attention is drawn to a strange division: The southern hemisphere of Mars is mostly rugged and intensely cratered, while the northern third of the globe is smooth, remarkably flat, and partially covered by vast lava flows. Geologists don't yet understand what caused Mars to end up looking like the halves of two very different planets, one very old and the other young, joined roughly along the equator.

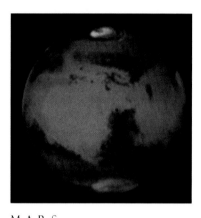

MARS

DIAMETER	4,220 MI
MASS	0.11 X EARTH
ROTATION PERIOD	24.6 HOURS
SURFACE TEMPERATURE	40°F
REVOLUTION PERIOD	1.88 YEARS

More of the landscape rolls by, and off to the south you spy an enormous depression. Known as Hellas, it is more than 2,400 miles across and 6 miles deep—the largest impact basin in the solar system. Hellas resulted from the collision of an asteroid or comet at least a hundred miles across. Many other large craters passing by below have been named for famous telescopic astronomers throughout modern history, including Herschel, Huygens, Cassini, and Lowell.

The terrain below begins to slope down gradually, leading to something very unexpected: a series of gigantic riverbeds, some tens of miles wide, winding their way around teardrop-shaped islands and northward toward the broad, flat lowland known as Chryse Planitia, the "Plain of Gold." Clearly, these were not normal waterways, the kind that on Earth gently guide meandering flows to the sea. Instead, the Martian channels' breadth and depth indicate that the water they once bore came down from the heavily cratered highlands in catastrophic floods, immense discharges that yielded a thousand to ten thousand times the flow of the mighty Mississippi. A few comparable events have occurred on Earth—for example, about 10,000 years ago the equivalent of a hundred Mississippis burst from a glacier-dammed lake and gouged out the Channeled Scablands of eastern Washington. But only on Mars do geologists find so many flooding episodes converging on a single basin.

Some of the channels lead upstream to the mouth of a gigantic complex of canyons, collectively called the Valles Marineris. Their immense scale easily dwarfs anything comparable on Earth: 2,500 miles long overall, with a central section 400 miles wide and 4 miles deep. Valles Marineris was not incised into the landscape like the puny Grand Canyon back on Earth. Instead, it is mostly a fault-formed gash, a line of weakness straddling the equator where the planet's crust ripped open billions of years ago. In some spots landslides have cascaded down the steep slopes to the valley floors.

As the canyonlands trace westward, the land around them begins to rise in elevation, finally cresting as a broad plateau named Tharsis. Comparable to Africa in extent and standing roughly five miles higher than its surroundings, Tharsis is the supercontinent of Mars. Geophysicists believe that an upward surge of molten rock within the mantle created this giant welt while the planet was still young. The brittle crust buckled and fractured from the strain, and great eruptions of lava surged across the landscape. To maintain the plateau's high standing, either the interior forces are still pushing upward or the crust has instead annealed and grown thick enough to support the great mass now topping it.

Either way, the Tharsis bulge dominates an entire hemisphere of Mars, and in turn five enormous volcanoes dominate Tharsis. They have the broad bases and gentle slopes characteristic of the shield volcanoes on the Hawaiian Islands, and each is capped with a broad caldera, or collapsed crater. Three of the five, named Arsia Mons, Pavonis Mons, and Ascraeus Mons, are spaced neatly along the crest line and rise more than 50,000 feet above it. At the plateau's north end is Alba Patera, a broad volcano comparable to Mount Everest in height but whose lava flows splay out for several hundred miles in every direction.

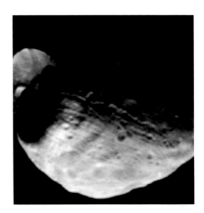

PHOBOS

DIAMETER	16 X 12 MI
MASS	0.00000015 X MOON
ROTATION PERIOD	7.7 HOURS
SURFACE TEMPERATURE	25°F
REVOLUTION PERIOD	7.7 HOURS

The undisputed king of Martian volcanoes, however, is Olympus Mons. Situated off the western flank of Tharsis, this immense mound of lava covers an area the size of Arizona and towers an incredible 75,000 feet above the lowland plains to its north. The caldera at its summit, a nest of several collapsed craters, is itself 60 miles across—big enough to hold metropolitan Los Angeles. Clouds often envelop the mountain's slopes and summit, making it one of the few Martian features recognizable from Earth. For more than a century, telescopic observers have known it as Nix Olympica ("Snows of Olympus").

LONG-AGO MARS. With all these impressive displays of geology, Mars has certainly had an exciting history. But most of the large features on its surface were shaped eons ago, when the interior still glowed from the throes of formation. Like all the other inner planets, Mars was initially hot enough to segregate into a crust, mantle, and core. The planet's iron-rich heart, somewhere between 1,600 and 2,100 miles across, once surged with electrical currents that generated a strong magnetic field. However, by the time Hellas formed, some four billion years ago, the field had died away. Why so? Mars is not nearly as dense as Earth, Venus, or Mercury, so perhaps the planet ended up with too little iron and heat-producing radioisotopes to sustain a molten core for very long. Or perhaps its smaller size allowed heat to radiate easily to space, quenching the internal fires early on.

Whatever the reason, we see little on the Martian surface that could be called "young," geologically speaking. The heavily cratered southern hemisphere clearly dates to very early in Martian history, and even late-appearing features like Tharsis and its towering volcanoes probably arose at least two and a half or three billion years ago. It is difficult to know for sure because we have no way to calibrate the landscape's crater populations, whether sparse or densely packed, with laboratory-deduced ages extracted from the rocks themselves. No clever machines or hardy astronauts have snatched samples of the red planet for study here on Earth.

Many astronomers think the Martian moons Deimos (lower left) and Phobos (lower right) were originally asteroids, like Gaspra (top). These three bodies do have similar shapes and sizes, though the surface of 7-by-10-mile Deimos has few large impact craters.

However, Martian rocks have come to *us,* free of charge, in the form of meteorites. The realization that pieces of Mars (and the Moon) lie scattered on Earth came to light in the early 1980s. At first, dynamicists couldn't imagine how an impact event could blast chunks of the Martian surface to escape velocity, three miles per second, without pulverizing them. But their geochemist colleagues insisted that a handful of meteorites match the chemical and isotopic signatures found by the Viking landers. Moreover, the lava-like fragments had solidified much too recently, between 150 million and 1.3 billion years ago, to have come from the Moon or any of the asteroids. Mars was the only viable candidate.

These meteorites tell us that volcanoes have erupted somewhere on the red planet in the not-too-distant geologic past. And in fact spacecraft images do show a few youthful features on the Martian surface, like fresh-looking lava flows around the base of the large northern volcano Elysium Mons. But the meteorites could be from most anywhere on the planet, leaving geologists to guess whether the Martian interior still churns with molten vitality or whether those recent volcanic eruptions are the final sputterings of a near-dead world.

Even more puzzling is what might be called the "Case of the Absent Aquifer." Water has clearly played a role in sculpting the Martian landscape. Besides the giant flood channels, the now-dry paths of many smaller streams can be traced among the craters and plains. Considering the evidence from just the giant floods alone, Mars has—or had—enough water to create a global ocean 2,000 feet deep.

So where did all that water go? Quite a lot has probably been lost to space over the eons, gradually stripped away by the ceaseless sweeping of the solar wind. Much more must be stashed in the planet's polar caps, dazzling white whorls that stand in stark contrast to the dark, ruddy plains around them. During winter, each cap grows larger as carbon dioxide from the already thin atmosphere freezes out and blankets more of the polar terrain. With the arrival of spring, the seasonal veneer of dry ice warms and vaporizes, revealing a residual slab full of water ice. These two polar reservoirs stand high and broad, and together they have sequestered roughly a million cubic miles of ice, only enough to make a modest 70-foot global layer.

The remainder must be stashed in the Martian crust, though any water near the surface is now solidly frozen as a globe-girding layer of buried ice. Subterranean ice is probably more concentrated closer to the poles, because craters well away from the equator often look as though they were splatted into mud. Elsewhere, river channels sometimes begin from chaotic jumbles—as if the permafrost suddenly melted and rushed out in a torrent, triggering wholesale collapse of the overlying ground.

Planetary scientists wonder if Mars was always cold and bleak, or whether it once enjoyed balmier times with a denser atmosphere and scattered seas. The landscape displays some tantalizing suggestions of a very different past climate. For example, hundreds of craters have fan-shaped deltas and layers of sediment in their floors, a strong indication that they once brimmed with water. In theory, as it grew to planethood, Mars should have acquired abundant carbon dioxide along with all that water. A dense atmosphere of CO_2 and water vapor would have provided a strong greenhouse effect, concentrating enough solar energy near the surface to raise the temperature by many tens of degrees.

But two key observations argue against a warm, wet early Mars. First, under those conditions rain and snowfall should have been common events. Yet, except for the giant flood channels, the landscape bears little direct evidence of precipitation: there are few small river systems and fewer still obvious signs of surface runoff. Second, standing water readily combines with CO_2 to form carbonate minerals. (Earth has little carbon dioxide in its atmosphere and a great deal of limestone on its seafloor for just this reason.) However, spacecraft scrutiny shows that no large carbonate deposits lie exposed on the Martian surface.

To complicate matters, some researchers have raised the possibility that a great sea once covered much of the planet's northern hemisphere. As evidence, they point to faint bands or terraces that may have served as shorelines. Moreover, the great northern plains remain very flat over thousands of miles, in some places sloping less than one foot per mile—a situation hard to explain by geologic processes alone. Fracture patterns in some of the lowest areas bear a striking resemblance to giant mudcracks and may indicate where the ground bowed upward after a massive overlying sheet of ice disappeared. Not everyone agrees that an ocean ever existed on Mars, because the ancient shorelines are vague at best. But if it did, the water inundated an area about one-third the size of the Atlantic Ocean and had an average depth of about 2,000 feet. Then this putative sea disappeared, probably three and a half to four billion years ago.

Planetary scientists don't yet understand long-ago Mars, nor can they say when water last gurgled across its landscape. In some places water may still be seeping out of crater rims and canyon walls, creating occasional muddy trickles down the steep slopes. Every 100,000 years, the Martian polar regions tip more directly toward the Sun, and it's unclear what climatic upheavals might occur at such times. Without question, the red planet remains as provocative now as ever. Venus may be Earth's twin in size and mass, but eventually we may come to realize that ruddy little Mars has most closely paralleled the evolution of our planet at ground level.

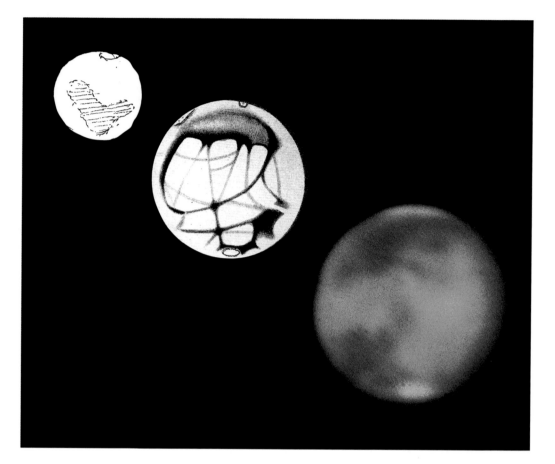

♂ MARS

Mars has long intrigued astronomers, tantalizing them with the shadowy features that cross its face. In 1672 the keen-eyed Dutch observer Christiaan Huygens identified the dark "sea" of Syrtis Major ("L" shape, upper left) and one of the planet's polar caps. Giovanni Schiaparelli saw a Mars streaked with *canali*—Italian for "channels"—in the late 1800s (center). Interpreted by some as the handiwork of an alien race, this web of waterways proved illusory. A telescopic photograph from 1969 (lower right) highlights the planet's variegated plains and its polar ice caps.

The idea that Mars held vegetation was not totally dispelled until the 1960s. By then cosmonauts, in spacesuits like this (opposite), and astronauts were racing to the Moon.

FOLLOWING PAGES: Twilight on Earth illuminates Mars-like terrain: Haughton Crater on Canada's Devon Island.

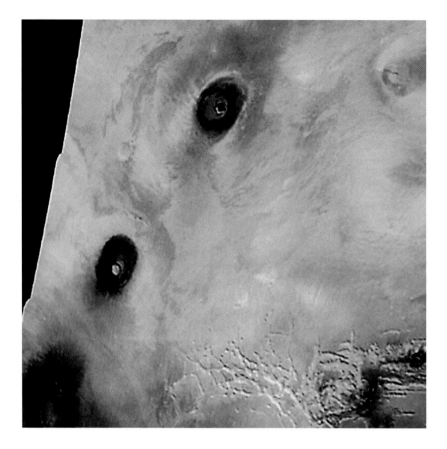

VOLCANOES The dark summits of
three enormous volcanoes—Arsia Mons, Pavonis
Mons, and Ascraeus Mons (above, top to lower
left)—jut 9 to 11 miles above the adjacent
Martian plains. They lie along the crest of the
planet's only "continent," a broad plateau named
Tharsis. The crisscrossing fractures at lower
right, Noctis Labyrinthus, mark the western
end of a giant canyon system.

An even larger volcano, Olympus Mons
(opposite), covers an area the size of Arizona.
At its summit, which towers more than 14 miles
above its surroundings, is a 60-mile-wide nest of
craters, collapsed into the mountain's eruptive
throat. Olympus Mons and its siblings resemble
shield volcanoes on Earth, like those of the
Hawaiian Islands. They probably grew to such
sizes at a time when the interior of Mars was
much hotter, and the volcanoes provided an easy
means of escape for the heat trapped within.

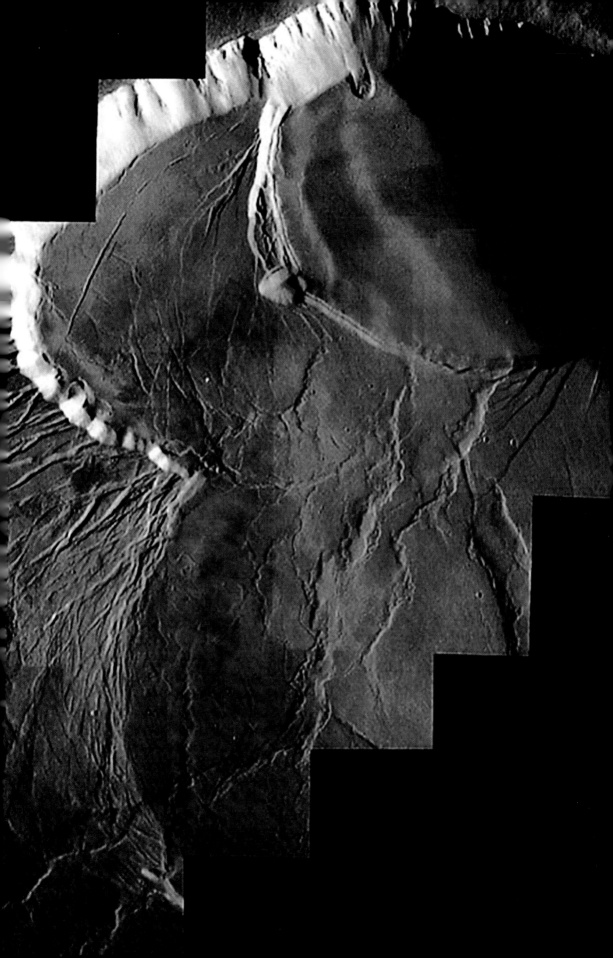

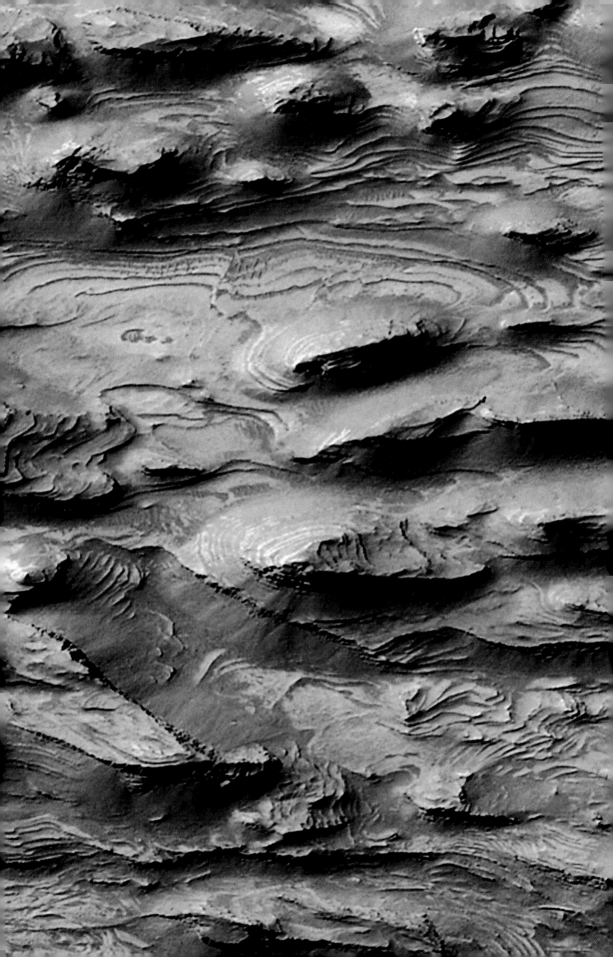

R O C K S Spacecraft have been circling and landing upon Mars for more than three decades, and their findings have revolutionized our knowledge of the Martian landscape. Layers of rock, stacked like pancakes on the floor of the Valles Marineris canyon system (opposite), were spied from orbit by the keen electronic eyes of NASA's Mars Global Surveyor. These may be multiple lava flows or sediments that settled to the bottom of an ancient lake.

Robotic landers have shown us a Mars that looks eerily reminiscent of the rocky deserts of Earth. In 1976 two identical Viking spacecraft used their mechanical arms (left) to scrape and scoop the sandy "soil" within reach. Mars Pathfinder (bottom, dark panel) and its roving sidekick, Sojourner, proved to be a hit, 21 years later, with scientists and the public alike as they reconnoitered a rock-strewn floodplain, largely unchanged for the last two billion years.

ODD ROCKS Although Mars is relatively stable now, its surface has been shaped over the eons by the erosive power of wind and water. Consequently, planetary geologists were bound to turn up a few oddities among the tens of thousands of pictures they have amassed of the planet's surface. Near Mars's south pole lies a stark, heart-shaped mesa (above). Less than 300 yards across, it is all that remains of a layer that once mantled all of the surrounding plains.

Another set of erosional survivors is found on the broad plain called Cydonia. Early reconnaissance by Viking I (below) revealed one high-standing mound that bears an uncanny resemblance to a human face. Although dismissed as a natural formation by geologists, some people speculated that "the face" was instead a calling card, carved by an ancient Martian civilization and left behind for Earthlings to eventually find. Decades later, new pictures taken with better cameras and different lighting confirmed that Mother Nature alone had chiseled the plateau's features. Such tricks of light and shadow abound on Mars. For example, aren't those bushes dotting the Martian hillsides (opposite)? That's certainly the impression given by this two-mile-wide scene. In reality, however, Mars Global Surveyor captured a field of sand dunes that had begun to lose their coating of carbon-dioxide (dry ice) frost—one triangular spot at a time—after a long, dark winter.

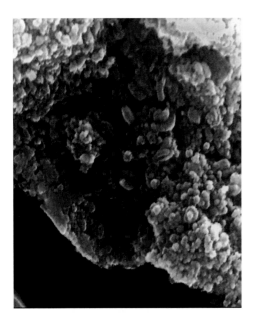

LIFE ON MARS Scientists now realize that some of the meteorites falling from the sky have been blasted off Mars. One of these (above), a four-and-a-half-billion-year-old specimen discovered in 1984, revealed clusters of tiny, elongated features when examined with a scanning electron microscope. Could these be the fossilized remains of ancient Martian microbes? Scientists still argue about whether this rock bears the fingerprint of life. Most are skeptical, but the evidence remains inconclusive.

A Martian "sol," or day, is only slightly longer than Earth's, so Mars Pathfinder watched the Sun rise and set dozens of times during 1997 from its landing site at Ares Vallis. Because the red planet's atmosphere is thin and dusty, the twilight sky takes on unusual colors: blue near the Sun and peachy pink well away from it. The spacecraft's camera also occasionally glimpsed clouds of ice crystals riding the winds at high altitudes.

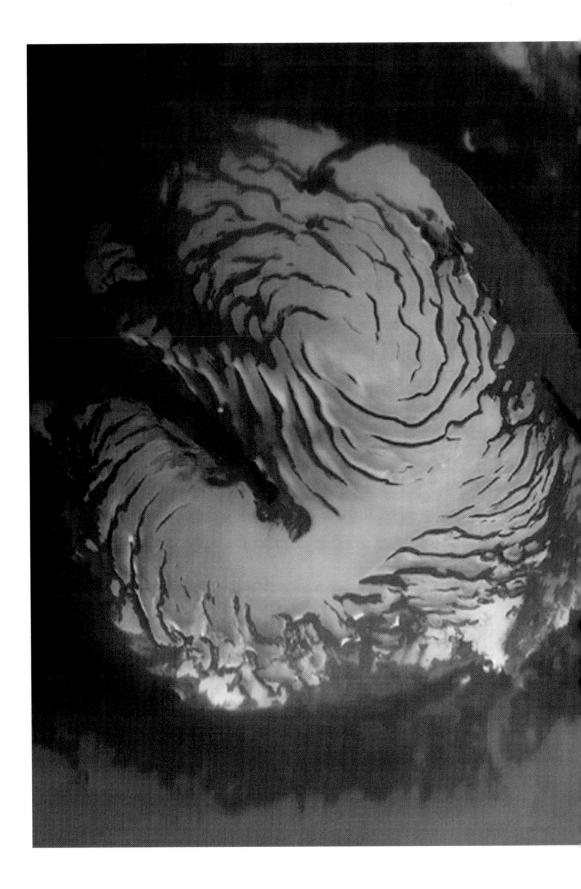

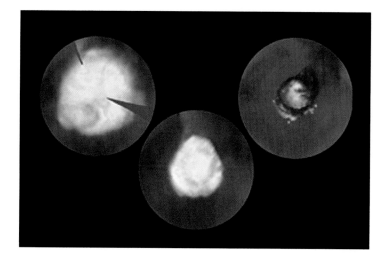

S T O R M S Giant pinwheels of ice and frost, the Martian polar caps contain a mix of frozen water and carbon dioxide. Because the red planet undergoes huge climatic swings every 100,000 years, thin layers of dust and ice create cyclic layers in the polar regions akin to tree rings. Some 700 miles across, the northern cap (left) is surrounded by a vast sea of wind-sculpted sand dunes.

From Earth, we can't look directly onto the Martian polar regions. But such scenes can be created (above) by combining snapshots from the Hubble Space Telescope. Using this technique, researchers watched the evolution of the Red Planet's northern pole over six months. In early spring the pole, left, was still cloaked with a heavy winter coat as an orange-hued dust storm blew across it. Its frost line had retreated by late spring, middle, and had shrunk to its minimum size by summer, right.

FOLLOWIING PAGES: Scores of gullies, likely cut by rivulets of water, cascade down a crater wall.

Dust clouds from a polar storm on Mars

A Saharan sandstorm escaping to the Atlantic

Storm of the Century

In May 1971, as Mars and Earth drew close together in their orbits, both the United States and the Soviet Union were ready to ratchet up the stakes in the exploration of the red planet. Soviet rocketeers dispatched two massive craft, named Mars 2 and 3, each designed to go into orbit around the planet and drop a camera-equipped lander onto the surface. The United States countered with two smaller Mariner spacecraft designed to orbit Mars for a comprehensive look at the planet's surface and its seasonal climatic changes. But Mariner 8 was lost during launch, leaving its twin, Mariner 9, to carry out all the mission's objectives.

Four months later, however, as the spacecraft sailed outward, Mars was whipping itself into a frenzy. It was summer in the planet's southern hemisphere, and the Sun's warmth had triggered strong winds and a growing dust storm. From Earth, astronomers watched as Mars's entire atmosphere became choked with ocher dust within weeks. Never before had they seen the planet enveloped in such an intense, long-lasting dust storm, and it couldn't have come at a worse time.

Mariner 9 reached the red planet first, firing its braking rocket and slipping into orbit on November 19th. Mars 2 arrived about two weeks later, separating into its lander (which crashed due to a malfunction) and a successful orbiter. Mars 3's lander reached the surface intact, but its transmissions ceased abruptly seconds after touchdown. No one knows why, but many believe the raging global dust storm may have toppled or damaged the lander. Unfortunately, because the craft's activities were completely preprogrammed and irreversible, Soviet engineers could not delay the landing until after the winds subsided. For the same reason, the Mars 2 and 3 orbiters automatically relayed to Earth a steady stream of pictures showing Mars's virtually blank disk.

Meanwhile, NASA engineers had the luxury of waiting out the storm, which at its peak extended an opaque shroud of dust to a height of more than 40 miles. Most of Mariner 9's scientific work was deferred until the atmosphere cleared, but the craft's early images showed a quartet of dark spots poking up through the dusty pall. Disbelieving scientists finally realized these had to be enormous mountains—crater-topped volcanoes, as it turned out—jutting from the surface. "I simply couldn't believe that they were volcanic," admits Bruce C. Murray, a planetary geologist on the project team. "In fact, not only volcanic, but larger than any comparable volcanoes on the Earth."

As the dust settled, Mariner 9 commenced its mapping duties. The camera returned pictures of the entire Martian surface and produced the first full-disk photo mosaic of Mars. By the following October, its attitude-control gas exhausted and its historic mission over, Mariner 9 had relayed more than 7,300 images to Earth. Even though a stuck filter prevented the camera from recording in color, the spacecraft nonetheless had surveyed the entire planet for the first time. It showed us a Mars than no one had imagined, a planet of towering mountains, vast canyon systems, intricately layered polar regions, and a network of desiccated riverbeds proving that water once flowed across the surface. After that, our notions of Mars as a planet, and as a target for future exploration, were forever changed.

MERCURY
& VENUS

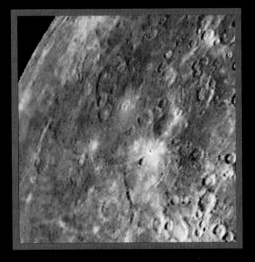

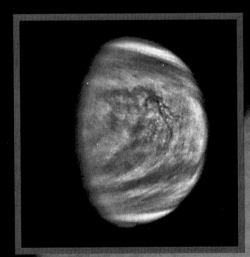

MERCURY & VENUS

Whenever astronomers turn to computers to simulate the formation of our solar system, they use the same basic approach: Start with hundreds (or thousands) of virtual planetoids in various orbits, then let them zip around the Sun, crashing into one another until a few large aggregations emerge victorious. Despite all the chaos, these planet-building exercises usually create a handful of rocky inner planets. Frequently one or two of them end up close to the Sun, and, in our own solar system, Mercury and Venus are those worlds.

Skywatchers have known about Mercury for thousands of years, a surprisingly bright, steady beacon that can be spotted with a little care in twilight before dawn or after sunset. But Mercury's small diameter (3,030 miles), and the fact that it never ventures farther than 28° from the Sun in the sky, makes this innermost planet very difficult to study telescopically. Not until 1964, thanks to advances in radar techniques, were astronomers finally able to determine Mercury's 58.6-day rotation rate and its rate of spin: exactly one and a half times per 88-day orbit. This arrangement means a hypothetical astronaut there would see the Sun rise only every 176 days, during which time the planet itself would have circled our star twice and spun completely around three times!

Only with the 1974 arrival of Mariner 10 did we get a close look at Mercury's surface. Thanks to fortunate orbital circumstances, the spacecraft was able to visit the planet three separate times as it looped around the Sun. Unfortunately, the same hemisphere was in sunlight each time, so despite three flybys Mariner 10 photographed only half the globe.

At first glance, Mercury looks remarkably Moonlike, with impact craters everywhere and occasional expanses of lava-covered plains. Much of the northern hemisphere is dominated by Caloris Basin, a large, multiringed impact some 830 miles across. The 100-mile-wide object that blasted out this huge scar struck about 3.85 billion years ago and created upheaval on a global scale. At the point on Mercury's globe exactly opposite ground zero, Mariner 10 photographed a tract of disheveled terrain unlike any other in the solar system. Its jumble of hills and fractures covers an area comparable to France and Germany combined. Many geophysicists suspect that Caloris's powerful shock wave propagated through and around the planet, coming to a climactic focus 180° away. The concentrated seismic energy caused the ground to lurch thousands of feet upward, and its collapse moments later left the region in chaos.

Although big, splashy impacts like Caloris are visually dramatic, it's the

PRECEDING PAGES: Cratered Mercury (top) and cloudy Venus; Mercury's silhouette on the Sun.
OPPOSITE: Cratered and outwardly Moonlike, the planet Mercury is not well understood.

small-scale details of the Mercurian landscape that continue to intrigue planetary geologists. For example, here and there the spacecraft recorded long, snaking faults where one slab of crust has ridden up over another. This hints that the planet was once molten to great depth, and as it cooled the planet shrank enough in overall volume that the outer crust (which would have solidified first) became compressed and fractured like the wrinkled skin of a drying prune.

MERCURY

DIAMETER	3,030 MI
MASS	0.06 X EARTH
ROTATION PERIOD	58.6 DAYS
SURFACE TEMPERATURE	800°F
REVOLUTION PERIOD	88 DAYS

Mercury's overall density, 5.4 grams per cubic centimeter, nearly equals that of Earth. So the planet must contain a great deal of iron, most of it concentrated in a core that takes up more than half of the interior. That big iron heart is probably responsible for the magnetic field detected by Mariner 10 during its flybys. But having a lot of iron does not guarantee magnetism. To generate a magnetic dynamo, Mercury's core must be at least partially molten, even though in theory it should have solidified almost completely by now, and the molten iron must be circulating, despite the planet's very slow rotation. Because the field is relatively weak, it could be literally frozen into a solid iron core (geophysicists term this remanent magnetism). But the true source remains unknown.

Since Mariner 10's visits, Earth-based telescopes have glimpsed a few features on the hemisphere unseen by the spacecraft. There's also a thin atmosphere, consisting of oxygen, sodium, and potassium vapor (likely launched into space by tiny meteoroids striking the rocky surface), along with helium and hydrogen (contributed by the solar wind).

Without question, the most remarkable and unexpected discovery came in 1991, when radar probing revealed that craters at the planet's poles are filled with some kind of radar-reflective deposit. Based on its radar properties, that material is most likely—and least logically—frozen water. When Mercury is closest to the Sun, its daytime surface soars to 800°F. However, many crater floors near the poles lie in permanent shadow and could, theorists say, provide safe havens for the water ice periodically brought to the planet by colliding comets.

Given Mercury's many mysteries, NASA plans to dispatch its Messenger spacecraft to the planet in 2004. Messenger will first fly past the planet twice at close range to probe its atmosphere and magnetic field. Then, in 2009, it will settle into an orbit for a year of detailed mapping and surface analyses. About that time the European Space Agency will launch its BepiColombo mission, whose two orbiters and small lander should reach Mercury after a two-and-a-half year flight.

In the years until these spacecraft arrive, astronomers will continue to explore this enigmatic planet of extremes with radar and other telescopes.

VENUS UNVEILED. Early telescopic observers had no better luck discerning the surface of Venus than they'd had with Mercury, but for a very different reason. The planet is completely enveloped in opaque white clouds, which show the vaguest hints of detail only when viewed in ultraviolet light. But even sight unseen Venus seemed much like Earth, a near twin in diameter (7,520 miles) and only 27 percent closer to the Sun. So scientific speculation about our "sister world" ran wild, conjuring up hypothetical landscapes that ran the gamut from molten puddles of zinc and lead to windswept deserts blanketing the globe, and from miles-thick polar ice caps to swampy tropical forests teeming with life.

The first clue that Venus was not simply a warmer version of Earth came in 1962, when radar soundings showed that the planet rotates very slowly, and in the direction opposite (retrograde) that of most other planets. That same year Mariner 2 passed 22,000 miles from Venus, close enough to determine that its surface was hundreds of degrees hotter than Earth's and that its atmosphere was both much denser and dominated by carbon dioxide. However, even though radar studies suggested the existence of several large, rough-sloped mountain regions, we still had little idea what the planet's surface looked like.

Fortunately, Venus is a relatively easy interplanetary target, and during the 1960s and 1970s a total of 23 Soviet and American spacecraft called upon Earth's veiled neighbor. We learned that the surface temperature of Venus has rocketed to 870°F and that its atmosphere exerts 90 times more pressure than that at sea level on Earth. Venus could

VENUS

DIAMETER	7,520 MI
MASS	0.81 X EARTH
ROTATION PERIOD	243 DAYS
SURFACE TEMPERATURE	870°F
REVOLUTION PERIOD	225 DAYS

not have gotten so hot just from being closer to the Sun. Instead, the hellish conditions result from an extreme version of the now-familiar "greenhouse effect." That is, the Sun's visible light can filter through the clouds to ground level, but the atmosphere's carbon dioxide does not let infrared (heat) radiation pass back out. So the heat becomes trapped near the ground, more than doubling the temperature that would exist there otherwise.

The first glimpses of the Venusian surface came in 1975 from the Soviet Union's Venera 9 and 10 landers, which succumbed to the infernal heat about an hour after relaying panoramas of a landscape littered with flat, rough-edged rocks. Three years later NASA's Pioneer Venus spacecraft arrived, a mission that combined a radar-equipped orbiter with a quartet of instrumented descent probes. The orbiter's surface map revealed a few highland plateaus amid extensive lowland plains, while the probes determined that the dense clouds consisted primarily of concentrated sulfuric acid. Venus was now a decidedly unpleasant place to visit! Even so, during the 1980s the Soviet Union continued its successful Venera series with four more landers that showed extensive lava flows on the surface and with two radar-equipped orbiters that mapped many new details in the planet's northern hemisphere.

In 1985 two Soviet craft, bound for Halley's comet, dropped off landing modules as they swung past Venus. As each landing craft fell toward the surface, it released a clutch of tiny instruments suspended from a balloon 11 feet across. The buoyant bubbles settled to an altitude of 34 miles, where their varnish-covered Teflon skins resisted the caustic sulfuric-acid clouds surrounding them. Whipped along by 150-mile-per-hour winds, the balloons traveled some 7,000 miles —a third of the globe's circumference—before their batteries failed.

From a geologist's perspective, however, the real breakthrough at Venus came in August 1990, with the arrival of NASA's Magellan orbiter and its sophisticated radar imaging system. The radar mapping progressed slowly but methodically, each orbit's data arriving at Earth as an image "noodle" revealing an area about 12 miles wide and nearly 10,000 miles long. By the time its mission ended four years later, the spacecraft had surveyed 98 percent of what lay hidden beneath the clouds, recording surface features down to stadium-size details.

Within weeks after the mapping effort began, Magellan's scientists realized that Venus is a world overrun by volcanism. They had expected to see some evidence of past eruptions, but nothing on this scale. Lava flows are everywhere. Small volcanic domes cluster together by the hundreds. In some spots, eruptions have produced enormous lava pancakes 40 miles across. Wherever the ground has fractured, rivers of molten rock have frequently streamed forth.

Another big surprise was the paucity of impact craters, especially small ones. Venus's atmosphere is so dense that it effectively shields the surface from any incoming projectiles less than 100 feet across. Slightly larger objects apparently break up during their atmospheric passage, sometimes creating misshapen clusters of overlapping pits when they strike the surface. Larger craters are often surrounded by contorted debris fields, suggesting that the dense lower atmosphere exerts considerable aerodynamic control over the material flying outward from the blasts.

The complete tally of Venusian craters falls just shy of 1,000, far fewer than would be expected for a surface presumably billions of years old. Moreover, the craters appear to be scattered randomly around the globe, indicating that the entire surface is about the same age. By extrapolating from the crater populations on the Moon, whose age is now known with certainty, geologists calculate that most (if not all) of the planet's exterior was emplaced no more than about 500 million years ago. R. Stephen Saunders, who headed Magellan's science team, notes that volcanic flows may be burying Venus's craters nearly as fast as they form. But it seems to him more likely that some geologic event triggered vast outpourings of lava. Only about 10 percent of the planet, the heavily fractured highlands, appear to predate the magmatic flood.

"What is amazing about Venus is that volcanic and tectonic resurfacing is not an ongoing process," notes Saunders. Instead, it seems to have run rampant for a short time, obliterating 90 percent of the planet's history in one geologic stroke, then gone into quiescence. Unfortunately, Saunders laments, this wholesale obliteration has robbed us of the opportunity to understand why and when the evolutionary paths of Venus and Earth diverged. But for now, after centuries of wondering, he and other planetary scientists are content to have seen the face of a planet that was once completely hidden from view.

OPPOSITE: A mile-high volcano (upper left) looms over equatorial Venus. Such computerized images, together with vertical exaggeration, help geologists understand the planet's evolution.

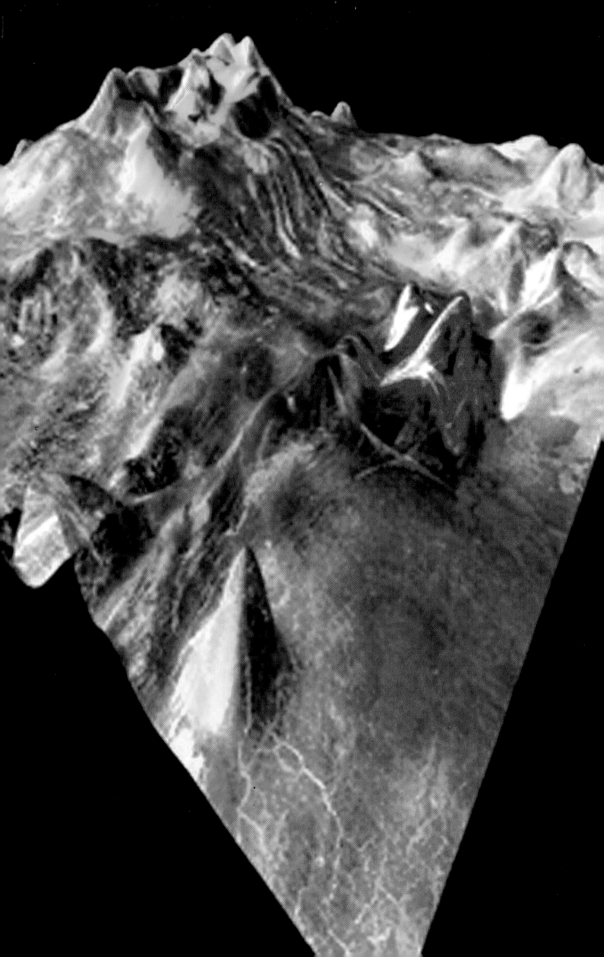

A Radar Race to Venus

With the dawn of the Space Age in the late 1950s came the realization that radar held great potential as an astronomical tool.

A successful radar contact with Venus would offer two key rewards. First, the planet could be used as a mileage marker to determine the Earth-Sun distance—a fundamental cosmic yardstick also known as the astronomical unit—to unprecedented precision. And astronomers would finally be able to probe the planet's surface through its clouds and determine its rotation rate.

The race was on, as three teams in the United States, plus others in England and the Soviet Union, used giant antennas to transmit powerful beams of radio energy to Venus and timed the round-trip echo time. To have any chance of success, these experiments had to be conducted when Venus and Earth were close together in space, an orbital geometry that repeats every 19 months.

The first radar contacts with Venus came during the 1961 viewing opportunity, which led to a revised value of almost 92,957,000 miles for the astronomical unit. But deducing the planet's rotation rate proved tricky, and initial results varied wildly from 20 to 300 days.

When Venus and Earth were once again close together, teams from the Jet Propulsion Laboratory in California and MIT's Lincoln Laboratory in Massachusetts were ready. In October 1962, using a recently built spacecraft-tracking antenna in the Mojave Desert, a team led by Roland L. Carpenter of JPL spotted a large, radar-bright feature on Venus's surface. He tracked its motion and concluded that

Venus was spinning every 225 days—but in the reverse, or retrograde, direction! (The true rotation period, we now know, is 243 days.)

Meanwhile, back in Massachusetts William B. Smith had already come to the same realization. Smith had approached

Paul Green, his supervisor at Lincoln Laboratory, for the go-ahead to publish the amazing result.

But Green balked. The lab's radar team had been embarrassed by trumpeting some bad data a few years before, and, in a decision he would later regret, Green insisted that Smith's unlikely result be watered down before publishing it.

Ultimately, credit for discovering Venus's odd, slow spin went not to Smith but to Carpenter and fellow JPL astronomer Richard M. Goldstein, who published their results in 1963.

Wearing 3-D glasses, researchers scan the Venusian landscape.

FOLLOWING PAGES: Over three weeks, Mercury makes its fleeting arch across the evening twilight.

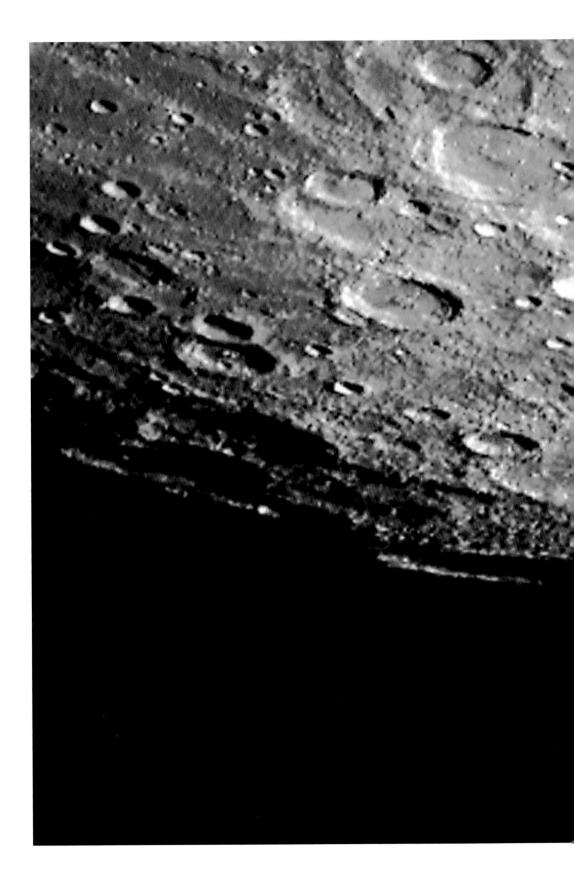

OTHER WORLDS

☿ MERCURY

Mercury's southern pole (opposite) looks much like its northern pole—and elsewhere: cratered and desolate. Although there are hints that this infernal planet has had a complex geologic history, much of its landscape remains unseen. To date our only close-up views have come from a single spacecraft, Mariner 10, which paid three brief visits in the early 1970s. The spacecraft's camera work, though highly prized, covered only half of Mercury's surface. Follow-up missions are planned for early in the 21st century.

A crack in Mercury's crustal armor (below)—a compression fault nearly 200 miles long—marks where one surface slab was thrust over another. Apparently the entire exterior of Mercury was once molten; as it cooled from the outside in, the planet shrank slightly. Geologists believe this contraction occurred about 3.9 billion years ago.

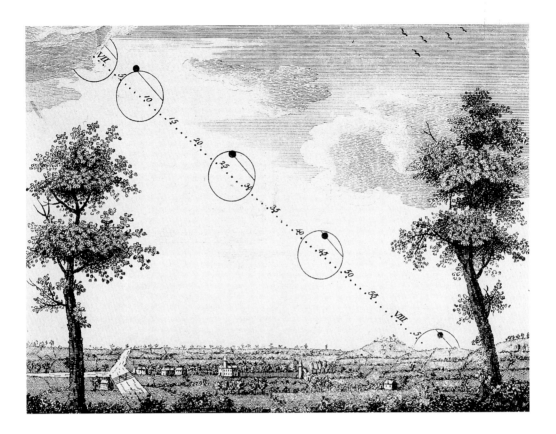

 # VENUS

Venus and Mercury are the only two planets that can pass directly between the Earth and Sun. Such transits were eagerly awaited spectacles, for by timing them carefully astronomers could determine the Earth-Sun distance very accurately. Transits of Venus, like that visible from London in 1769, depicted above, are rarer than those involving Mercury. They occur in pairs, eight years apart, yet only a dozen or so take place in a span of a thousand years. Mark your calendars: The next ones are in 2004 and 2012.

Venus and Jupiter, the two brightest planets, pass just a degree or so from each other as seen through the slit of the Gemini North telescope in Hawaii. (The facility is so named because it has a nearly identical twin 6,600 miles to the southeast, in Chile.) Seen through a telescope, Venus mimics the whole range of lunar phases: It can appear as a thin crescent, as gibbous, or as "full."

FEATURES Computer wizardry transports us (above) to some four miles over the Venusian landscape. Below we behold the large volcano named Sapas Mons. Hundreds of miles away, beyond a vast lava plain, towers five-mile-high Maat Mons, another of Venus's thousands of volcanoes. This virtual sightseeing is better than being there. Were we actually to brave Venus's atmosphere, we'd choke on the thick clouds of sulfuric acid and be incinerated by the hellish temperatures. These conditions meant a brief life for the Soviet Union's Venera spacecraft, eight of which dropped onto the planet's surface in the 1970s and early 1980s. Even the hardiest of them lasted only a few hours, though four took pictures of the stark terrain surrounding their landing sites. In the panorama radioed to Earth by Venera 13 (right), the cloud-filtered sunlight imparts an orange cast. The sawtooth-edged ring is part of the craft's landing mechanism.

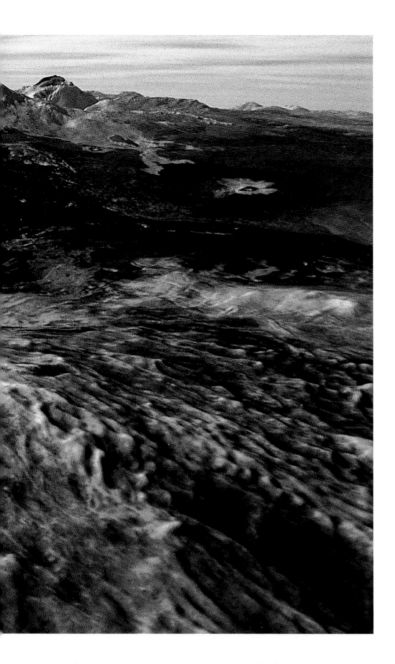

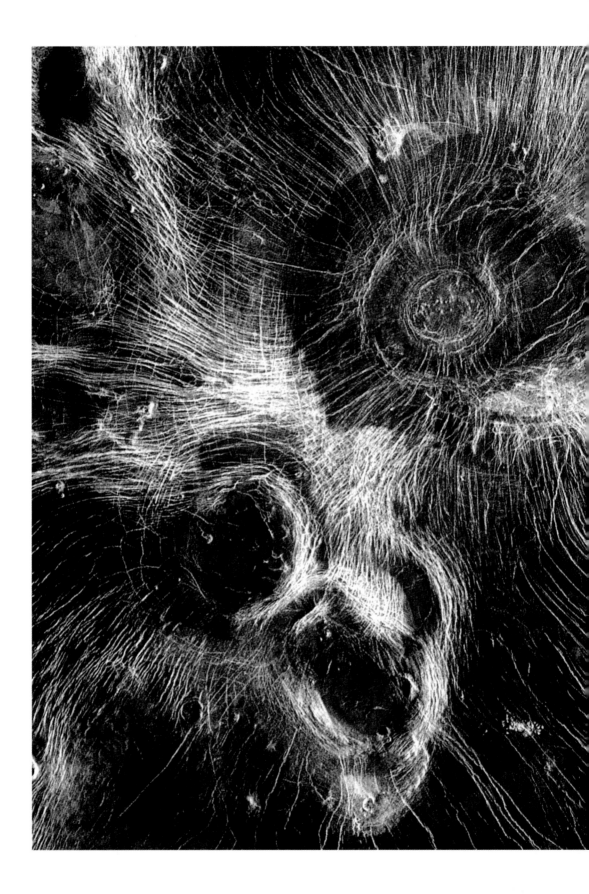

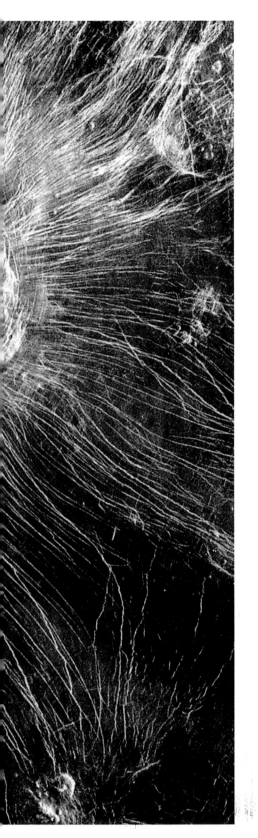

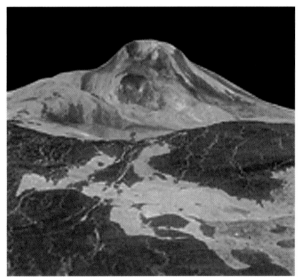

VOLCANOES Thousands of volcanoes are scattered across Venus. One of the most prominent is five-mile-high Maat Mons (above), which looks imposing in this vertically exaggerated perspective. Lava flows, which appear bright to Magellan's radar imager, extend for hundreds of miles across the fractured plains seen in the foreground.

Orbiting high above Venus, the Magellan spacecraft used its cloud-penetrating radar to record a volcanic wonderland. More than 80 percent of the planet is covered with lava flows, lava-capped plains, and other volcanic features like these "arachnoids" (left), which are surrounded by a web of fractures. Similar to other circular features called coronae, arachnoids most likely form when magma rises to just below the surface, causing the ground to bulge and fracture. Then the magma chamber cools and retreats, and the ground above it collapses.

HOME PLANET

HOME PLANET

I t's a good bet that we humans have wondered what lies beneath our feet since prehistoric times. Seven hundred years ago, Dante's epic *Inferno* described nine descending strata—increasingly harsh levels of hell—leading to Satan himself at Earth's core. In 1665 Athanasius Kircher, a Jesuit scholar, proposed a subterranean realm crisscrossed by tubes of lava and hot air. Such Hadean concepts were perhaps not so far-fetched after all, in light of the huge volumes of molten rock sporadically seen erupting out of the ground. In fact, to understand Earth you must first appreciate the dominant role that heat has played in its evolution.

Think of our planet's formation four and a half billion years ago not as a gentle sweeping up of solid matter from the solar nebula but instead as a ferocious, 100-million-year-long assault. Every infalling planetoid created a pulse of heat when it struck Earth's rapidly growing mass, and eventually this onslaught rendered the outer shell completely molten to a depth of several hundred miles or perhaps much more. Viewed from space, our planet was a giant glowing ball surrounded by an atmosphere of superheated steam and carbon dioxide.

Within the deep ocean of magma, light and heavy elements began to segregate from one another like oil from water. Great volumes of iron sank toward the center, a colossal falling out that itself raised the planet's temperature by thousands of degrees. Minerals rich in silicon, calcium, and aluminum floated to the top, creating a low-density silicate "froth" that congealed into the crust. It was at this point, most geochemists believe, well along in the assembly process, that an object at least the size of Mars struck the nascent Earth and splashed out enough matter to form the Moon.

By the time the bombardment died away about four billion years ago, outwardly Earth looked solid, but surrounding its massive core—a highly compressed ball of solid iron—was a thick sphere of molten metal. Churning motions in this outer, liquid core created electric currents, the dynamo responsible for Earth's magnetic field. Within the overlying mantle and crust, radioactive elements began to decay, releasing even more heat. Much of this radiated away directly into space while the crust remained thin. But the balance of energy changed once it thickened, and, like a house with extra insulation, the interior heat's easy escape route was largely cut off.

CONTINENTS ADRIFT. Given their similar size and density, Earth might well have ended up like Venus, forced to cool down through ceaseless, random volcanism. Instead, somewhere between one and a half and three billion years ago,

PRECEDING PAGES: A thunderstorm, evening planet and Moon collide on a Kansas hilltop.
OPPOSITE: Mir space station and a natural satellite, the Moon, ride above a stormy Earth.

our planet developed a way to let off steam found nowhere else in the solar system: plate tectonism. Earth's rigid exterior broke into pieces, or plates, that began to slide atop the partially molten upper mantle. Wherever the plates spread apart, molten rock gushed out through the opening. Where they pushed together, one would override the other, forcing the loser down into the mantle like an ice cube thrust into a hot drink.

EARTH

DIAMETER	7,926 MI
MASS	81 X MOON
ROTATION PERIOD	23.93 HOURS
SURFACE TEMPERATURE	59°F
REVOLUTION PERIOD	1 YEAR

Despite four decades of intensive research, geophysicists cannot agree on why Earth developed plate tectonism in the first place. Perhaps it was the lubricating effect of the mingling of seawater with the upper mantle. Whatever the reason, this self-sustaining cycle is what keeps Earth's interior from overheating. Nor is there scientific consensus about what sustains the cyclic activity: Clearly, the plates move in response to stirring from below, but is the driving force created by the upper mantle or by buoyant plumes of magma rising from the core-mantle boundary?

Today Earth has eight large intersecting plates, along with many smaller ones, and if we could strip away all our planet's water and vegetation, their ragged-edged patchwork would be apparent. The largest ones underlie the Pacific Ocean, Eurasia, Africa, each of the Americas, and Australia. The continents, which are the thickest portions of Earth's crust, serve as the core sections of many plates. Deep offshore trenches and rugged mountain ranges mark where one crustal slab is subducting (being shoved under) another. The Pacific Ocean is disappearing virtually around its entire rim—from the Aleutian and Kuril Islands in the northwest to the Andean coast of South America. That boundary is known as the "Ring of Fire" because of its frequent volcanic eruptions and earthquakes.

All the plates are slowly but constantly moving, jostling, and rearranging themselves, which means that their locations must have been different in the distant past. The seeming fit of South America's eastward-jutting "hip" to Africa's west-coast "waist" is no coincidence. Roughly 120 million years ago they lay snugly against one another. But then a rift zone, the Mid-Atlantic ridge, cleaved them apart. Today the ridge is marked by a long snaking chain of volcano-lined fractures that, except for its dramatic slicing through Iceland, lies completely hidden beneath the Atlantic Ocean. Earlier still, about 260 million years ago, all of Earth's major landmasses were clumped together in one hemisphere, forming a supercontinent known as Pangaea. By fast-forwarding the plates' motion into the distant future, geophysicists predict that the Pacific Ocean basin will be smaller, the Atlantic larger, and Australia's northward migration may put it on a collision course with Japan.

Other planets show no evidence of plate tectonism, and Earth appears to be its only practitioner. Apparently the rigid exterior of Venus became too thick (and its rocks too dehydrated) to initiate any overturn. Smaller Mars, with less heat to dissipate and a greater surface area relative to its volume, easily radiated excess energy to space and may now be completely solid throughout. Mercury's situation remains unclear.

Another deceptively simple question: Why is our planet so wet? Earth formed in an environment too hot for water to condense directly from the solar nebula, so all that liquid must have been imported from elsewhere. One long-held view was that a swarm of comets entered the inner solar system roughly four billion years ago, delivering Earth's oceans through collisions. However, there's an important isotopic mismatch: The ratio of deuterium ("heavy hydrogen") to ordinary hydrogen in cometary water is about twice the ratio found in seawater. No more than one-tenth of Earth's water could have arrived via dirty snowballs without skewing the deuterium-to-hydrogen ratio too much. Moreover, there's no compelling explanation for how and why so many comets could have arrived in such a geologically brief time (roughly 100 million years).

Primitive asteroids, not comets, may have been the dominant source of Earth's oceans. Early in solar-system history the region now known as the asteroid belt teemed with planetary embryos and asteroids, many containing up to 20 percent water. Jupiter's early development caused the overwhelming majority of these bodies to be cleared out. Many of the belt's escapees should have collided with the nascent inner planets. By then Earth had become massive enough to keep most of the collisional ejecta from being lost to space, so the impactors contributed their considerable bulk to our planet's growth and supplied more than enough water to fill its oceans.

MOON

DIAMETER	2,160 MI
MASS	0.01 X EARTH
ROTATION PERIOD	27.3 DAYS
SURFACE TEMPERATURE	245°F
REVOLUTION PERIOD	27.3 DAYS

THE MOON AS COMPANION. Although Earth's Moon is not strictly a planet, its nearness to us makes it the obvious center of our nocturnal attention. Its 2,160-mile-wide face displays a combination of light and dark markings that ancient skywatchers imagined to consist of *terrae* (highlands) and *maria* (seas). Today we realize that these regions represent two distinct rock types, each billions of years old and each with much to tell us about the Moon's formation and evolution. From a scientific perspective, the lunar story is largely written in the 842 pounds of samples carried home by Apollo astronauts and the additional two-thirds of a pound returned by Soviet robotic landers.

The myriad craters that can be glimpsed through a telescope testify to the antiquity of the Moon's surface, but only after the Apollo expeditions could we peg the true age at four and a half billion years. Lunar history probably began with a bang, when a large protoplanet struck a glancing blow to the still-forming Earth and shot enough vaporized rock into orbit to yield a single large offspring. Molten and white hot at the outset, the Moon quickly segregated into layers, as Earth had. But the collision's ejected debris contained so little iron that the resulting lunar core can't be more than about 500 miles across, only a few percent of the total mass. The rock "froth" that floated to the top became the light-colored crust.

Violent impacts continued to reshape the Moon for another half billion years. The most destructive events fractured the lunar crust hundreds of miles below the surface, deep enough that molten lava could later gush from the interior and flood the freshly excavated basin floors, forming the maria. Virtually all of the eruptions died away by about 3.1 billion years ago, leaving our satellite much as it appears today. However, what we see is only half of the story, because the same hemisphere always faces inward. Put another way, the Moon takes the same amount of time (27.3 days) to spin on its axis and complete one orbit. It's a forced situation that arose long ago from the constant tugging by Earth's gravity on a not-quite-symmetrical Moon.

Jules Verne, the father of science fiction, imagined lunar exploration. In his novel *From Earth to the Moon,* first published 1865, three heroes are encased in a 20,000-pound aluminum projectile and fired from a cannon.

Lacking an atmosphere or a magnetic field, the Moon is completely exposed to the harsh realities of interplanetary space. For example, over time impacts have reduced its outer layer to complete rubble. In fact, virtually all of the Apollo samples are amalgams of many different rocks that have been fused together by the heat and shock of countless cratering events. Seen at the microscopic level, the lunar rocks and dust grains show incredibly tiny impact pits, mineral crystals damaged by cosmic rays, and surfaces impregnated with gases from the solar wind.

Amazingly, our Apollo samples show no trace of water or its influence whatsoever— the Moon's formation left it as dry as dust. Yet astronomers believe that caches of water ice exist near the lunar poles, buried in the floors of deep craters that never see the Sun. The ground in this permanent polar night hovers around -370°F, providing an ideal "cold trap" for any water brought to the lunar surface by comets. Just how much ice might exist is a guess, though it could be tens of millions of tons—enough to sustain future lunar colonists for many centuries.

Having such a large satellite has influenced our planet in two profound ways. Initially the Moon and Earth were *much* closer together, separated by no more than about 10,000 miles. A full Moon, in addition to being very dramatic, would have raised enormous tides in our early oceans. However, back then (as now) our planet's brisk rotation was dragging the tidal bulges somewhat ahead of the Earth-Moon line. Because the Moon is constantly pulling on those bulges, the slight misalignment has gradually slowed Earth's rotation—every century the day becomes 0.0014 second longer—and the Moon gets a little energy boost that pushes it more than an inch farther away from us each year.

The second effect involves the long-term stability of Earth's obliquity, or polar tilt. Right now our planet's polar axis tips toward the Sun by 23.5°, though over 41,000 years this value oscillates by 1.5° one way or the other. That may not seem like much of a change, but it's enough to cause the protracted periods of deep cold that trigger ice ages.

Things could be much worse, however. On Mars, which doesn't have a sizable satellite, the polar tilt swings through a wide range every 100,000 years. Without the steadying influence of the Moon, Earth's obliquity would make wild, erratic swings every few million years and, in extreme cases, could be anywhere from near 0° to as much as 85°. Thanks to the Moon, we don't have to worry about the drastic climatic and biological consequences of Earth's polar regions' abruptly shifting toward the Sun.

TARGET EARTH. There are other long-term concerns for our planet, however. Beginning in the 1970s, astronomers became keenly aware that asteroids and comets continue to strike the inner planets. Impact scars are particularly evident on the battered face of our Moon, where craters stand shoulder to shoulder over much of the near side and almost all of the far side. On Earth craters are much rarer—only 160 have been confirmed—because the erosive power of wind and water fill in or wear down most of them in just a few million years.

Calculating the odds of Earth getting clobbered are difficult, primarily because we have yet to discover all the big rocks floating in space that could potentially strike us. Astronomers have cataloged several hundred "near-Earth objects," or NEOs, that are at least a half mile across. Statistical studies imply that there are perhaps a thousand of these in all, which means that most have yet to be found. Millions more must exist at smaller sizes.

Even an object just 200 feet wide will do a great deal of regional damage when it hits. That's the size of whatever exploded over Siberia's Tunguska region in June 1908, flattening 800 square miles of uninhabited taiga. Such blasts occur every few centuries, whereas a mile-wide object strikes Earth once in a million years on average. And the most devastating impacts, from asteroids and comets at least five to ten miles across, lay waste to our planet every 100 million years. The last of these cataclysms was 65 million years ago, an event that gouged out a 120-mile-wide crater on what is now the Yucatán Peninsula and eradicated 70 percent of all species then alive.

Fortunately, the cosmic disasters portrayed in the blockbuster movies *Deep Impact* and *Armageddon* have yet to happen in modern times, and no known asteroid or comet is on a collision course with us. However, the emphasis here is on "known" — many potentially threatening objects lurk undiscovered in the interplanetary space surrounding Earth. Should astronomers identify such an asteroid, chances are good that its arrival will be many years or decades in the future. Comets collide with the inner planets less often, but they are harder to spot far in advance and their orbital motion less predictable. In any case, how we will deal with such a remote but terrifying eventuality remains unclear.

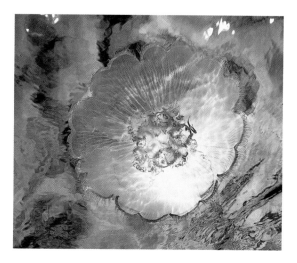

⊕ EARTH

Life on our planet takes on forms both strange and familiar, the consequence of eons of adaptation to the thin, near-surface veneer called the biosphere. A moon jelly (above) in the Tasman Sea, seen from below, is hardly distinguishable from the sky. Translucence proves a valuable defense from predators for such jellies, which are unable to flee and can seldom find shelter in the ocean. The mucus on top of their bells as well as short tentacles help moon jellies capture prey. Below this jelly is a temporary fish visitor.

On Earth, a small planet in a crowded solar system, we equate green with abundant life. Few locales can match the explosion of life found in a tropical forest, such as the Monteverde Cloud Forest Preserve in northern Costa Rica (left). Yet ours ultimately is a fragile existence. If it had no atmosphere, our planet would be an ice world with an average temperature near 0°F. Were it much closer to the Sun, greenhouse gases would run amok, forcing the oceans to evaporate.

W A T E R Iguaçu Falls, at the border of Brazil, Paraguay, and Argentina, serves to remind us that we are land dwellers on a wet planet. All told, Earth has some 400 million cubic miles of water, enough to create a layer two miles deep over the entire globe. Through tides, evaporation, precipitation, and runoff, water is central to the climate of Earth. Chemically speaking, however, we know that not all of it was present during our planet's formation. The ratio of hydrogen to deuterium (a heavier hydrogen isotope) in water tells geochemists that some of it—and perhaps most of it—arrived sometime later, borne here by a succession of comets and asteroids.

FOLLOWING PAGES: Most of the Earth's land has been scorched by fire at some time in the past.

PLATE TECTONISM

Our planet's crust is on the move, as typified by a computerized view looking along California's 800-mile-long San Andreas Fault (above). Derived from radar data, this perspective follows the fault as it trends southeast along the mountains in the Temblor Range near Bakersfield, California.

Using a technique called seismic tomography, akin to taking a CAT scan of the Earth's interior, geophysicists can now accurately gauge the temperatures and structures found deep within our planet. Computer models then provide an "insider's perspective" on how things work (right: red shows hot rock rising; blue shows cold rock sinking). Adam Dziewonski, a pioneer of seismic tomography, finds plenty of evidence that the lower mantle is constantly churning. His results reveal a number of cold slabs beneath the Pacific, subducted into the deep interior long ago, and a plume of hot, rising rock under the Hawaiian Island chain. So far as we know, Earth is the only planet that recycles crustal plates in order to cool its interior.

ASIA

AUSTRALIA

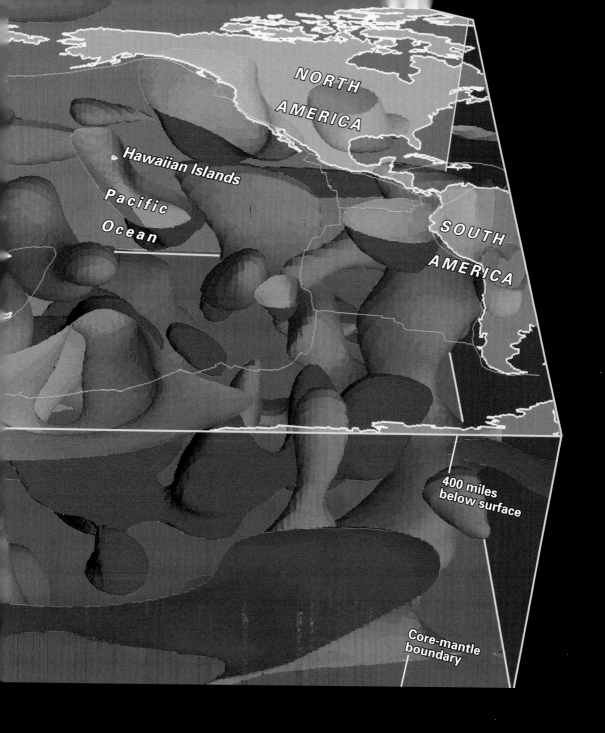

NORTH
AMERICA

Hawaiian Islands

Pacific

Ocean

SOUTH

AMERICA

400 miles
below surface

Core-mantle
boundary

VOLCANOES Rimmed by towering volcanoes like Mount Hood in Oregon (below) and Klyuchevskaya volcano (right) on Siberia's Kamchatka Peninsula, the Pacific Ocean's circumference is well known to geophysicists as the "Ring of Fire." Beneath the ocean waters lies a battle zone where huge slabs of crust collide and sideswipe one another. The oceanic plate, thinner and more pliable than the continental slabs that surround it, usually loses in head-on competition; it is forced beneath (subducted) the continental margins, resisting its plunge into the mantle with violent earthquakes and chronic volcanism. This image of Klyuchevskaya volcano was acquired by the Spaceborne Imaging Radar-C/X-band Synthetic Aperture Radar (SIR-C/X-SAR) aboard the space shuttle *Endeavour* on October 5, 1994. The volcano had begun to erupt on September 30. Kamchatka's volcanoes are among the most active in the world.

Earth Reconstructed

Scientists, only human, sometimes continue to cling to ideas even when the evidence for change is overwhelming. Geologists, for example, earned a reputation for stubbornness throughout much of the 19th and 20th centuries, especially when it came to understanding the global-scale evolution of our planet. Not until the 1960s did they come to realize—and begrudgingly accept—that Earth's crust is subdivided into a series of huge, mobile plates that have been pushing each other around for a billion years or more.

In retrospect, the hints were everywhere. For centuries cartographers had noticed that the Atlantic shorelines of Europe, Africa, and the Americas could be nested together like some giant jigsaw puzzle. The match was more than just a good fit: Large topographic features like mountain ranges could be stitched together as well, and fossils of ancient plants and animals from opposite sides of the Atlantic Ocean bore remarkable similarities. The world's paleontologists were also hard pressed to explain why the remains of so many tropical species were preserved in the frigid tracts of Greenland and Antarctica.

Alfred Wegener, a turn-of-the-century meteorologist, was hardly the first person to take note of all these oddities, but he became obsessed by them. Pulling together all the available evidence, in 1912 Wegener began to lecture about his ideas of "continental displacement," or what came to be called continental drift. In his model, our planet's landmasses initially lay clumped together over one of the Poles. Over time, Earth's rotation, combined with tidal forces from the Sun and Moon, pulled the conti-

nents apart and distributed them around the Equator. Wegener published his radical concepts in 1915, in the face of fierce criticism from most established geologists in Europe and, especially, the United States. Since the ocean floors consist of solid rock, they argued, there was no obvious way for the continents to bulldoze their way over the surface to reach their current locations.

Wegener died in 1930, his ideas relegated to the fringe of geologic thinking. But the continental coincidences would not go away, even as new evidence began to emerge. During the 1920s a German oceanographic expedition used sonar to discover that a long mountainous ridge runs down the middle of the Atlantic Ocean for thousands of miles. By the 1950s American geophysicists realized that a deep canyon ran through this underwater mountain range like some giant tear in Earth's crust. Moreover, the ocean floor to either side of the rift exhibited a pattern of mirror-image magnetic stripes caused by reversals in Earth's magnetic field, with the youngest stripes nearest the ridge's center.

All the geophysical signs pointed to a crust in constant motion, but how? The breakthrough came in 1960, when geologist Harry Hess proposed that the continental migration resulted from seafloor spreading. Although the mobilizing forces are not yet understood, the oceanic crust spreads apart at the mid-ocean ridges, pushing and jostling continents along the oceans' margins. The Americas, Europe, and Africa *did* once nest together after all, a gathering now known as Pangaea. Hess had validated Wegener's long-ignored claims, and the science of plate tectonics was born.

OPPOSITE: New crust (red) forms along a central ridge that runs the length of the Atlantic Ocean.

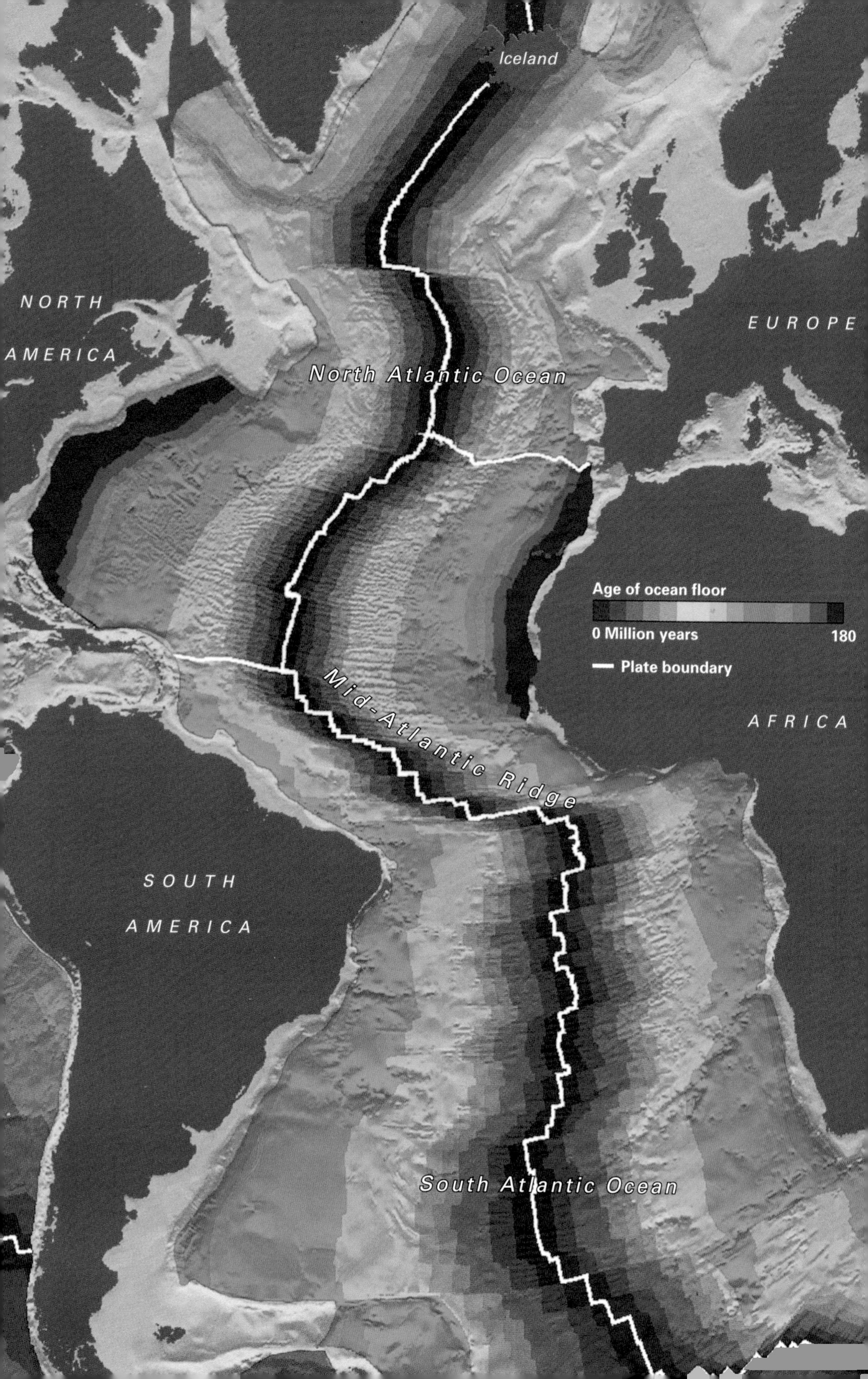

Iceland

NORTH

AMERICA

North Atlantic Ocean

EUROPE

Age of ocean floor

0 Million years   180

━━ Plate boundary

AFRICA

Mid-Atlantic Ridge

SOUTH

AMERICA

South Atlantic Ocean

Making the Moon

Who hasn't gazed at a slender lunar crescent delicately suspended in the evening sky, or watched with rapt awe as a full Moon rises majestically over a distant horizon? Each of us has surely pondered, at one time or another, how Earth came to have such a stunning satellite. Of course, astronomers have wondered too. In fact, determining the Moon's origin was a major reason for having Apollo astronauts trudge across the stark lunar landscape to scoop up its dusty rocks.

The truth is that we still don't know how the Moon formed—but we can now rule out some long-held speculations. Before the Apollo expeditions, theories of lunar origin were variations on three basic themes. The first held that the Moon formed in Earth's general vicinity and became a captive body when it ventured close to our planet. The second envisioned Earth and Moon forming together as a kind of double planet. And the third suggested that at some point the early Earth was spinning so rapidly that a large mass ripped away and into orbit.

Today none of these ideas stands up to close scientific scrutiny because any formation scenario must satisfy several rigorous physical and chemical constraints. For example, the Earth-Moon system has too much angular momentum—our planet spins too fast, and the Moon circles it too quickly—for the two bodies to have simply formed side by side. Conversely, Earth would have had to spin incredibly fast, in just 2.6 hours, for the Moon to have cleaved away, and the resulting pairing would have been left with four times too much angular momentum. Lunar samples returned by the Apollo and (Soviet) Luna programs show

that, unlike Earth, our cosmic companion is severely lacking in iron and certain other elements (tungsten, for example) that tend to aggregate wherever iron does. Nor do lunar rocks contain any water whatsoever. Yet they contain oxygen's three isotopes in ratios that are virtually identical to Earth's— in fact, the bulk lunar composition roughly matches that of our planet's mantle.

These harsh post-Apollo realities left scientists bereft of good ideas, save for one. In the mid-1970s, two teams independently suggested that the Moon could have been blasted away from the early Earth in a titanic collision. In Arizona, William K. Hartmann and Donald R. Davis concluded that collisions among the inner planets' precursors would have been common, just as Alastair G. W. Cameron and William R. Ward, in Massachusetts, realized that a big, glancing blow to Earth could create a large satellite and fortify the system with plenty of angular momentum.

In the years since, several computer modelers have simulated how such a collision would play out. They conclude that the impactor had to be something at least as massive as present-day Mars, and that before striking it must have already segregated into an iron-rich core and a less-dense mantle. An off-center hit would have sprayed a huge jet of iron-poor mantle material into space as white-hot vapor, while the impactor's core remained behind and merged into Earth's. The vapor quickly smeared out into a doughnut-shaped disk, and as it cooled, the gas would condense into a chaotic swarm of tiny particles.

Then a fight for survival broke out around a critical boundary at an altitude of 7,500

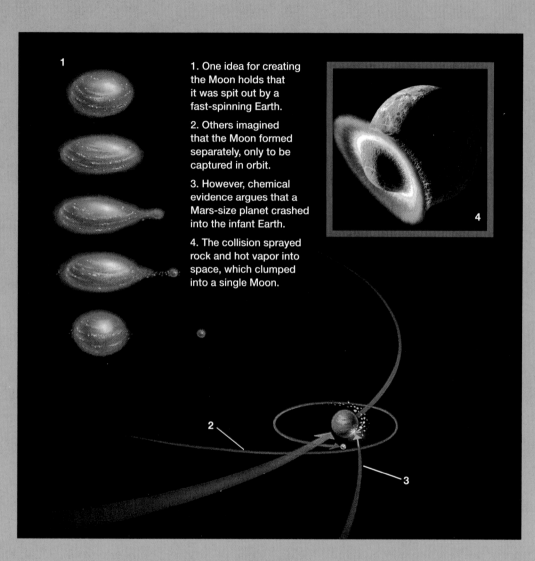

1. One idea for creating the Moon holds that it was spit out by a fast-spinning Earth.

2. Others imagined that the Moon formed separately, only to be captured in orbit.

3. However, chemical evidence argues that a Mars-size planet crashed into the infant Earth.

4. The collision sprayed rock and hot vapor into space, which clumped into a single Moon.

miles. Most of the disk cascaded back into the seething Earth, while higher-flying matter collected into ever-larger clumps. The simulations suggest that the Moon assembled in as little as one month. That timing appears to have been critical. Gravitational perturbations between the nascent Moon and the disk's remaining matter created a mutual repulsion, which pushed the still molten satellite outward and into an orbit tilted about 10° to Earth's Equator.

Impact specialists find this scenario quite appealing, and it neatly explains most of our satellite's major characteristics—its size, composition, and orbit—without having to resort to contorted logic or unlikely physics. All it took was one chance collision in the right place, at the right time.

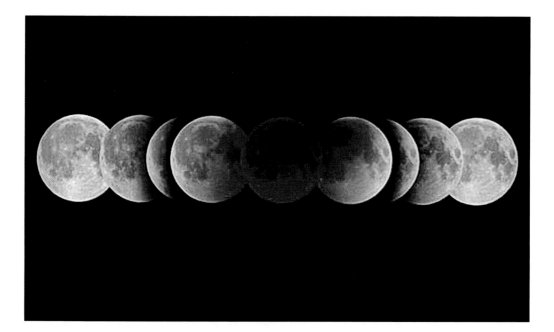

M O O N

During its monthly trek around Earth, the Moon sometimes slides through our planet's shadow, the umbra. A lunar eclipse in July 2000 (above) was widely observed from locales rimming the Pacific Ocean, and a composite of photographs records the Moon's right-to-left motion through darkness. At mid-eclipse, the feeble sunlight glancing around Earth's rim—the light of a thousand sunsets—gives the Moon a dramatic coppery cast.

The full Moon (opposite), seen through the electronic camera of the Galileo spacecraft in 1992, reveals the familiar face that has captivated human imagination since prehistoric times. Seventeenth-century scholars, thinking the Moon must be like Earth, concluded that the dark areas were *maria* (Latin for "seas"). Today we know that these are great plains of dark lava that gushed from the lunar interior for a billion years after the outer crust solidified. The bright-centered splash at bottom left is the crater Tycho.

FOLLOWING PAGES: Apollo 17 astronaut Harrison Schmitt confronts a pair of giant lunar boulders.

EPILOGUE: LIFE'S NICHES

Humans lead a sheltered existence. We cannot survive if the weather gets too cold or too hot, or if the oxygen in our air gets too thin. We have to be careful what kinds of plants we eat and to avoid too much (or too little) sunlight. Yet we are hardly representative of Earth's inhabitants—*Homo sapiens* may be at the top of the food chain, but by sheer numbers this is a bug's world. Very little is known about the vast microbial universe that exists beneath our feet, though biologists appreciate just how varied and robust these basic forms of life are. (The estimated total weight of all living bacteria is a hundred trillion tons, which, if spread evenly over all of Earth's continents, would make a layer five feet thick.)

A remarkable characteristic of a group of microbial species collectively known as extremeophiles is their tolerance of—and often their preference for—a wide range of harsh environments. One unlikely habitat is frigid Antarctica, where simple organisms inhabit ice-covered lakes and dry, windswept valleys. Another is the murky ocean bottom, where hot geothermal vents host a veritable oasis of microbial activity. Totally isolated from sunlight, organisms there thrive on the heat and nutrients gushing up from beneath the seafloor. Some of these microbes do best at or near the boiling point of water (one heat-loving species can't cope well at all below 194°F), or in extremely briny, acidic, or alkaline environments. They have even been found miles deep within the Earth itself, literally living off the bare rock around them.

Our planet teems with life today, but it wasn't always such a hospitable place. For hundreds of millions of years after it came together 4.5 billion years ago, Earth was too hot and too ravaged by impacts to sustain life. The pummeling from asteroids and comets ended about 3.9 billion years ago, perhaps punctuated by a devastating blitzkrieg at the end. Soon thereafter the Earth's surface cooled and oceans formed. Just exactly when life gained its foothold is a matter of considerable debate, however. As we attempt to peer further back in time, the biological signposts become less obvious. More complex life-forms like shell-shielded invertebrates appeared little more than a half billion years ago, and before then the fossil record consists mostly of mats of algae (called stromatolites) and other microscopic curiosities.

For a time researchers could trace the tree of life back no further than 3.5 billion years. Were this the true inception of all biota, it would mean that life took several hundred million years to take hold after the heavy cosmic bombardment ended. However, in 1996 an international team of researchers

OPPOSITE: Looking bizarre and menacing, long-horned beetles of French Guiana can be four inches long. Many insect species have earned Earth its reputation as a bug's world.

Resembling an "organic spaceship" in this image, the iridescent sea walnut shimmers with color, a lighting scheme perhaps designed to startle or confuse predators.

announced a remarkable finding. A set of banded, iron-rich rock formations on Greenland's Akilia Island yielded an excess of carbon-12, an unmistakable fingerprint of this planet's life (which prefers this isotope over less common carbon-13 for its cellular structures). Because the host rock is 3.85 billion years old, this discovery pushed the biological horizon back another 350 million years. Apparently, life arose on Earth rapidly, geologically speaking—perhaps in less time than Earth itself took to assemble from the chaos of the solar nebula—and it's been flourishing ever since.

In 1996 a team of scientists made the controversial claim that a four-and-a-half-billion-year-old meteorite from Mars, found 13,000 years after it landed in Antarctica, contained evidence of fossilized microbial life.

The implications of this rapid rise for finding life elsewhere are profound. No longer must we think in terms of a half-billion-year gestation, when instead a mere 50 million years might do. Nor do we necessarily need to confine our search for life-forms to idyllic tidal pools sheltered from harsh climates. Instead, we can look just about anywhere that organic compounds, water, and a suitable energy source are found together.

Water seems to be an essential prerequisite for life, no matter how extreme, but it is not available everywhere in the solar system. Earth lies in what is sometimes termed the zone of habitability, a narrow range of distances from the Sun at which water can exist as a liquid on a planet's surface. Were Earth just 15 million miles closer to the Sun, the few tens of degrees of added warmth would have disastrous effects. The oceans would evaporate more readily, charging the atmosphere with water vapor, thus enhancing the greenhouse effect and raising the temperature even more. Eventually the oceans would dry up and expose their vast seafloor sediments. Carbon dioxide from decomposing carbonates would accumulate rapidly and trigger a runaway greenhouse in the atmosphere, quickly sterilizing the entire Earth. Meanwhile, sunlight would break down water molecules in the upper atmosphere, allowing their hydrogen atoms to escape to space. Gradually, but irreversibly, our planet would desiccate.

Moving Earth outward from its current orbit would be equally bad. With more of its water frozen, rain would become a rare occurrence. Without frequent rainfall, the CO_2 gas from volcanoes would concentrate in the atmosphere. Somewhere near the orbit of Mars, or a bit farther out, dense clouds of carbon-dioxide ice crystals would cloak the planet and reflect away precious sunlight. Temperatures would plummet, sending Earth into a global deep freeze.

By good fortune, Earth appears to have achieved Goldilocks's delicate balance: not too hot, not too cold, just right. However, astronomers point out that 3.8 billion years ago, about when life first took hold here, the Sun was significantly fainter and shone with only 75 percent of the energy it pumps out today. Back then Earth's average surface temperature should have been about 60°F lower, well below the freezing point of water. Since the biological evidence tells us that our planet was not completely iced over throughout its youth, something must have moderated the temperature. Most likely, Earth's carbon dioxide had not yet been locked away in ocean sediments but resided in the atmosphere with a thousand times its present-day abundance. The strong greenhouse effect that resulted has staved off the big chill for more than four billion years.

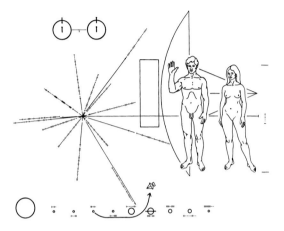

Earth lucked out on several counts, it seems, and today our planet abounds with life. Are there other inhabited planets, we wonder? For all of recorded history, and probably long before, our ancestors pondered whether life existed somewhere among the fixed and wandering

Now headed for the stars, two Pioneer spacecraft carry gold-anodized plaques, each of which bears an encrypted message designed to alert alien civilizations to the existence of intelligent life on planet Earth.

stars. In the late 1800s certain scientific circles considered the habitation of other planets very likely. At the turn of the 20th century the French Academy of Sciences offered a huge cash prize for anyone who could confirm the existence of life elsewhere—except on Mars, which the sponsors thought would be too easy.

BELIEVING IN THE EXISTENCE OF EXTRATERRESTRIAL LIFE is one thing; finding it is proving very difficult. From space, it's not obvious that Earth itself is populated. Scientists conducted a mock search for telltale biological clues when the Galileo spacecraft flew past Earth in 1990 en route to Jupiter. Its sensors found abnormal amounts of oxygen and methane in our atmosphere (two gases unlikely to be found together) and strangely discolored areas on the surface. But the real giveaway was the detection of single-frequency transmissions from our radio and radar systems.

Outwardly, the most likely abode of alien life is Mars. In 1976 two instrumented Viking landers set down on the planet and tested its soil for life. None was found, nor was any trace of organic matter. The planet's atmosphere is too thin to shield the surface from the sterilizing power of the Sun's ultraviolet light. But it's not hard to imagine a long-ago Mars with a denser, more protective atmosphere and a warmer, wetter climate. Given its rapid rise on Earth, life could have formed on the red planet's surface, then died off. Perhaps it took refuge underground (or arose there). Biologists point to the thick stacks of sediment seen ubiquitously across the Martian landscape as likely places to conduct future digs. In time, we will know if Mars has life now or ever did. Spacecraft are being designed to land there, retrieve samples of its surface, and return them to Earth—or to the quarantine of an orbiting space station— for sophisticated laboratory analysis.

In 1996 a research team from NASA and Stanford University claimed that a Martian meteorite contains remnants of tiny organisms that existed on the red planet billions of years ago. Using sophisticated microscopic techniques, they had found strange little shapes that look like the very smallest bacteria known on Earth. The rock also contains traces of organic molecules and other evidence that could be interpreted as being due to biologic processes. However, few of their colleagues have been swayed by the evidence—perhaps the strange shapes and compositions have nothing to do with biologic activity or betray contamination picked up during the meteorite's 13,000 years on Earth.

Another candidate abode is Europa, one of Jupiter's four largest moons. From the outside Europa's icy crust looks like a cracked eggshell, beneath which may lie an ocean of liquid water. The ocean would be kept from freezing by tidal friction created by Jupiter's strong gravitational pull, and it may lie very near the surface in places. No one knows whether sunlight is filtering into the briny depths, or whether the moon's interior is still warm. But it is tempting to draw comparisons between Europa's putative ocean and the hydrothermal ecosystems found astride Earth's mid-ocean ridges. Even if Europa is frozen throughout now, it may still have been mantled with a gently sloshing ocean throughout much of solar-system history.

The search for life's niches is no longer confined to our own solar system, because dozens of stars are known to have planets encircling them. For now, we cannot observe these worlds directly; they betray their presence by creating subtle but detectable wobbles in the motion of their host stars. Presumably these stars have their own habitable zones, and it's easy to imagine worlds with warm oceans. Answers will only come once a new generation of space telescopes can detect these alien worlds directly and provide hints of their character.

Until then, we remain the only planet known to harbor life. In trying to look at our own situation objectively, we are confronted with stark contradictions. On one hand, we find organisms living, even thriving, in extreme environments. Moreover, fossil evidence argues that life arose early in Earth's history (and, given the intense bombardment in primeval time, perhaps more than once). On the other hand, the entire biosphere is confined to a layer only one-thousandth of Earth's diameter. That thin veneer sustains life with ample liquid water, a palatable mixture of atmospheric gases, enough ozone to block the Sun's ultraviolet light, and a magnetic field to shield us from space radiation.

Whether life ultimately proves common or rare in the universe, our curiosity—about the worlds around us and about ourselves—compels us to probe ever farther for the answers. Our explorations have only just begun.

OPPOSITE: Life exists on Earth in extreme environments. A scanning-electron-microcope image reveals unidentified bacteria (colored red) that thrive in salty, highly radioactive conditions.

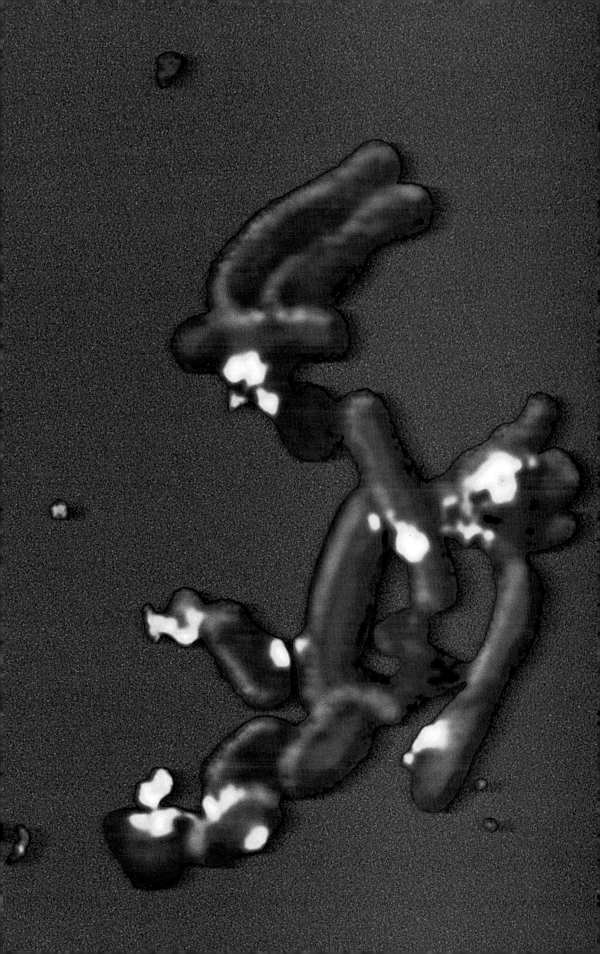

ADAPTATION

More than a mile and a half down in the Pacific Ocean, giant tube worms flourish in warm water bubbling through a 35-foot lava pillar. The worms escaped recent lava flows that buried other colonies off the west coast of Mexico in 1991. Returning to the eruption site three years later, researchers documented the phoenix-like rebirth of communities of organisms around the region's hydrothermal vents. Such rapid recovery is a hallmark of life, whether on land or under the sea.

Based on observations like these, biologists and paleontologists question the notion that life arose on Earth in primordial tidal pools warmed by the Sun and endowed with some kind of organic-rich "soup." Instead, they now give the inside track to deep-sea and subterranean environments completely cut off from sunlight. Satisfying their meager nutritional needs from water and simple organic compounds, the first microbes probably drew energy from the heat escaping our planet's interior or from chemical reactions. Such conditions are not unique to Earth.

PAST LIFE

Looking like a planet adrift in space, a tiny ball of carbon (right) found on what is now an island off Greenland holds a world of meaning as perhaps the oldest evidence of life on Earth. Colored here for contrast and magnified some 6,900 times, the ball nestles in a cavity etched from rock that is 3.86 billion years old. Despite the specimen's having lost all its anatomical features, scientists think its biochemistry must have been similar to that of all life that has evolved since.

Over the eons, life on Earth may have waxed and waned, but a quarter of a billion years ago, at what is called the Permian extinction, it came precariously close to being snuffed out entirely. At that time, long before the emergence of mammals, the ten-foot predator Dinogorgon hunted on floodplains in the heart of today's South Africa. In less than a million years Dinogorgon vanished (above) in the greatest mass extinction ever, along with about nine of every ten plant and animal species on the planet.

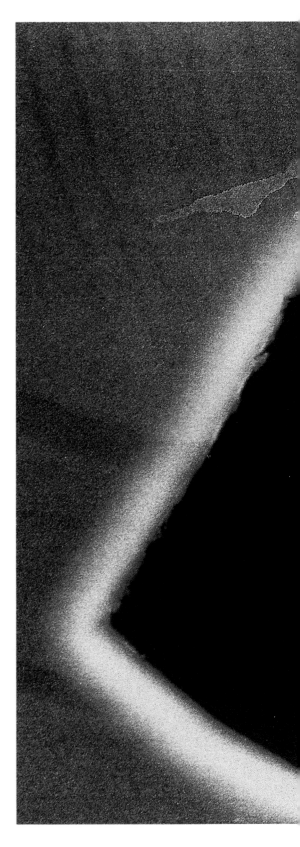

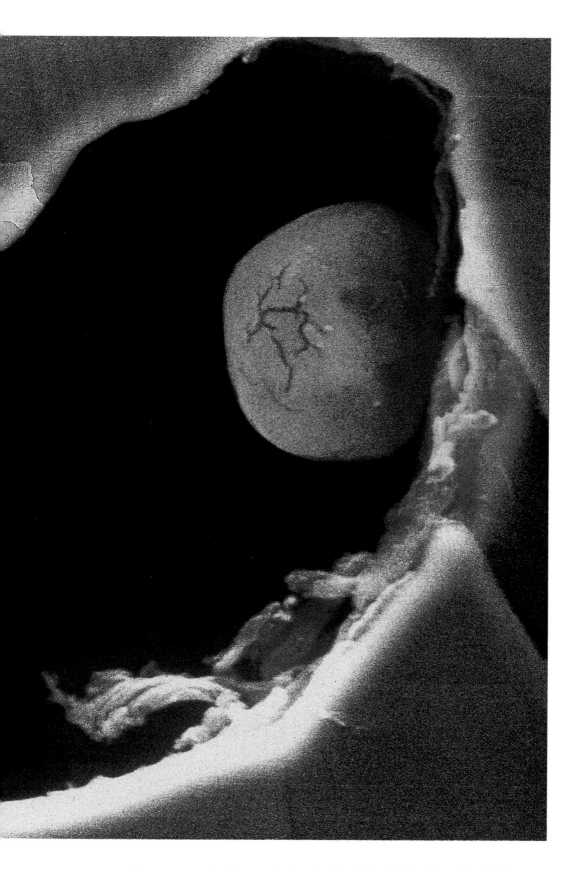

FOLLOWING PAGES: A universe can be imagined in a whole-sky infrared map taken by the Cosmic Background Explorer (COBE) satellite. The blue S-shape is dust in our solar system.

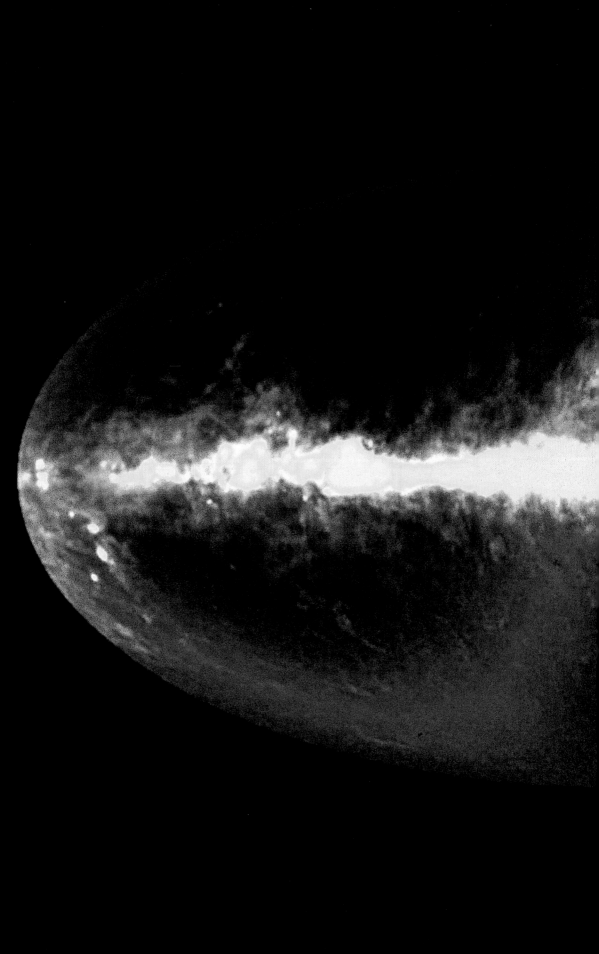

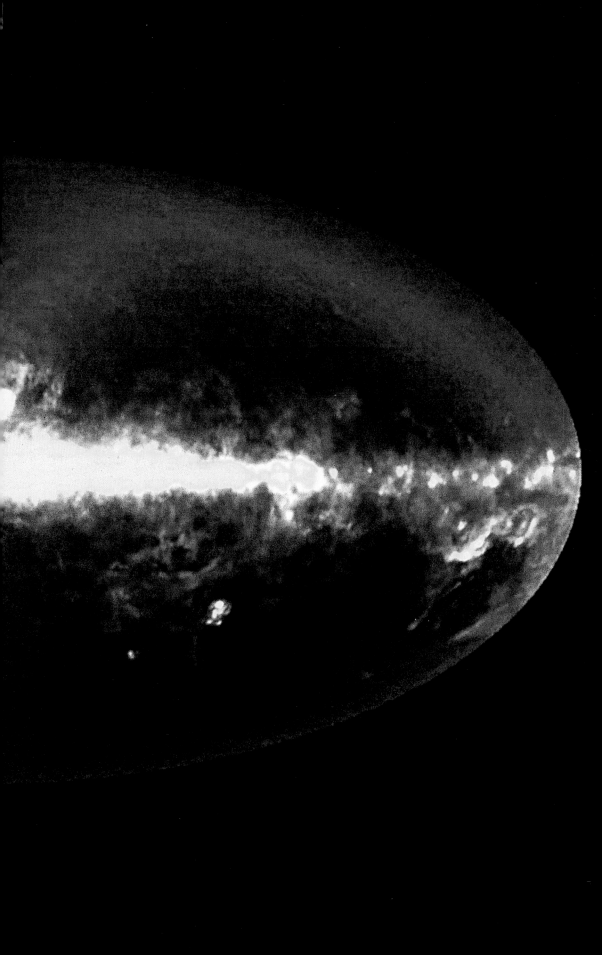

ABOUT THE AUTHOR

J. Kelly Beatty has spent more than a quarter century writing about the solar system and space exploration. He joined the staff of *Sky & Telescope* magazine in 1974 and now serves as its executive editor. He conceived and edited *The New Solar System*, which is considered a standard reference among planetary scientists. Beatty holds a bachelor's degree in geology from the California Institute of Technology and a master's degree in science journalism from Boston University. Asteroid 2925 Beatty was named on the occasion of his marriage in 1983.

AUTHOR'S ACKNOWLEDGMENTS

"No man is an island," and no book of this scope ever reaches its intended readers without the efforts of talented individuals who work to make it a reality: Barbara Payne and Tom Melham who helped lay the foundation for the book; project manager and illustrations editor Greta Arnold, who set the stage with her captivating illustration selections and kept the project focused and on track; art director Cinda Rose, who created a handsome design package; text editor Marty Christian, who ensured that only the best words got into print; researcher Gary Roush, who combed through a mountain of references to keep me honest; text editors Rebecca Lescaze and Lyn Clement, who read final stages of proof; and Melissa Ryan, who expertly consulted on digital imagery retrieval and management.

Finally, writing this book would not have been possible without the support of my wife, Cheryl, who never whimpered despite all those lost nights and weekends.

FURTHER ACKNOWLEDGMENTS

The National Geographic Book Division wishes to thank the following scientists for consulting on the book: Don L. Anderson, California Institute of Technology; Alan P. Boss, Carnegie Institution of Washington; Peter G. Brown, Los Alamos National Laboratory; Robin M. Canup, Southwest Research Institute, Boulder, Colorado; G. Edward Danielson, California Institute of Technology; James Garvin, NASA; Owen Gingerich, Harvard-Smithsonian Center for Astrophysics; Richard M. Goldstein, Jet Propulsion Laboratory; James W. Head III, Brown University; William B. Hubbard, University of Arizona; Bruce C. Jakosky, University of Colorado; Linda Kelly, the Planetary Society; David L. Levy, Jarnac Observatory; Mark Littmann, University of Tennessee, Knoxville; Jonathan I. Lunine, University of Arizona; Jeffrey M. Moore, NASA/Ames Research Center; Mark Robinson, Northwestern University; Fred L. Whipple, Harvard-Smithsonian Center for Astrophysics; and Donald K. Yeomans, Jet Propulsion Laboratory, Pasadena, California.

ADDITIONAL READING

Beatty, J. K., and others, eds., *The New Solar System* (Sky Publishing and Cambridge University Press, 1999); Ferris, T., *Life Beyond Earth* (Simon & Schuster, 2001); Jones, B. W., *Discovering the Solar System* (John Wiley & Sons, 1999); McSween, Jr., H. Y., *Fanfare for Earth: The Origins of Our Planet and Life* (St. Martin's Press, 1997); Norton, O. R., *Rocks from Space: Meteorites and Meteorite Hunters* (Mountain Press Publishing, 1998); Raeburn, P., Mars: *Uncovering the Secrets of the Red Planet* (National Geographic, 1998); Shirley, J. H., and R. W. Fairbridge, eds., *Encyclopedia of Planetary Sciences* (Chapman & Hall, 1997); Stern, A., and J. Mitton, *Pluto and Charon: Ice Worlds on the Ragged Edge of the Solar System* (John Wiley and Sons, 1998); Wood, J. A., *The Solar System* (Prentice Hall, 1999).

ILLUSTRATIONS CREDITS

INDEX

Boldface indicates illustrations.

Exploring the Solar System
Other Worlds

J. Kelly Beatty

Published by the National Geographic Society
John M. Fahey, Jr., *President and Chief Executive Officer*
Gilbert M. Grosvenor, *Chairman of the Board*
Nina D. Hoffman, *Executive Vice President*

Prepared by the Book Division
Kevin Mulroy, *Vice President and Editor-in-Chief*
Charles Kogod, *Illustrations Director*
Barbara A. Payne, *Editorial Director*
Marianne R. Koszorus, *Design Director*

Staff for this Book
Greta Arnold, *Project Editor and Illustrations Editor*
Martha Crawford Christian, *Text Editor*
Cinda Rose, *Art Director*
Gary W. Roush, *Researcher*
R. Gary Colbert, *Production Director*
Lewis R. Bassford *Production Project Manager*
Sharon Kocsis Berry, *Illustrations Assistant*
Julia Marshall, *Indexer*

Manufacturing and Quality Control
George V. White, *Director*
John T. Dunn, *Associate Director*
Vincent P. Ryan, *Manager*
Phillip L. Schlosser, *Financial Analyst*

Library of Congress Cataloging-in-Publication Data

Beatty, J. Kelly.
    Exploring the solar system : other worlds / J. Kelly Beatty.
        p.   cm.
    Includes bibliographical references and index.
    ISBN 0-7922-7878-X (reg.) — ISBN 0-7922-7875-5 (dlx.)
        1. Solar system. I. Title.

    QB501 .B4 2001
    523.2—dc21   2001034252

Composition for this book by the National Geographic Society Book Division. Printed and bound by R. R. Donnelley & Sons, Willard, Ohio. Color separations by Quad Graphics, Martinsburg, West Virginia. Dust jacket printed by Miken Companies Inc., Cheektowaga, New York.